**ACPL ITEM
DISCARDED**

3-14-77

Manufacturing Processes: Ceramics

MODULAR EXPLORATION OF TECHNOLOGY SERIES

ENERGY AND TRANSPORTATION: INDUSTRY AND CAREERS
John J. Geil and B. Stephen Johnson

ENERGY AND TRANSPORTATION: POWER
John J. Geil

ENERGY AND TRANSPORTATION: AUTOMECHANICS
Clifford E. Simes

ENERGY AND TRANSPORTATION: SMALL ENGINES
James N. Yadon

COMMUNICATION: INDUSTRY AND CAREERS
Rex Miller

COMMUNICATION: ELECTRICITY AND ELECTRONICS
Rex Miller

COMMUNICATION: GRAPHIC ARTS
Darvey E. Carlsen and Vernon A. Tryon

COMMUNICATION: DRAFTING
Thomas J. Morrisey

COMMUNICATION: PHOTOGRAPHY
Philip G. Gerace and Stephen S. Mangione

MANUFACTURING: INDUSTRY AND CAREERS
Harvey R. Dean

MANUFACTURING PROCESSES: WOODS
Gerald D. Cheek

MANUFACTURING PROCESSES: METALS
H. C. Kazanas and Lyman Hannah

MANUFACTURING PROCESSES: PLASTICS
Donald J. Jambro

MANUFACTURING PROCESSES: CERAMICS
Thomas G. Gregor

CONSTRUCTION: INDUSTRY AND CAREERS
William P. Spence

CONSTRUCTION: ARCHITECTURAL DRAWING
William P. Spence

CONSTRUCTION: SYSTEMS AND MATERIALS
William J. Hornung

CONSTRUCTION: TECHNIQUES
Glenn E. Baker

CAREERS: AN OVERVIEW
Robert M. Worthington

INTRODUCTION TO DESIGN
Robert Steinen

Modular Exploration of Technology Series

Manufacturing Processes: Ceramics

Thomas G. Gregor

Temple University
Philadelphia, Pennsylvania

Consulting Editor
H. C. Kazanas

PRENTICE-HALL, INC., Englewood Cliffs, New Jersey

This book is dedicated to Julia, George, and Bob Gregor

Library of Congress Cataloging in Publication Data

Gregor, Thomas G   1944–
    Manufacturing processes—ceramics.

    (Modular exploration of technology series)
    Includes index.
    1. Ceramics.   I. Title.
TP807.G773   666   75-43656
ISBN 0-13-555672-4
ISBN 0-13-555664-3 pbk.

MODULAR EXPLORATION OF TECHNOLOGY SERIES

Manufacturing Processes: Ceramics

Thomas G. Gregor

© 1976 by Prentice-Hall, Inc., Englewood Cliffs, New Jersey 07632. All rights reserved. No part of this book may be reproduced in any form or by any means without permission in writing from the publisher. Printed in the United States of America.

10 9 8 7 6 5 4 3 2 1

PRENTICE-HALL INTERNATIONAL, INC., London
PRENTICE-HALL OF AUSTRALIA, PTY. LTD., Sydney
PRENTICE-HALL OF CANADA, LTD., Toronto
PRENTICE-HALL OF INDIA PRIVATE LTD., New Delhi
PRENTICE-HALL OF JAPAN, INC., Tokyo

Modular Exploration of Technology Series

This book is one module in the MODULAR EXPLORATION OF TECHNOLOGY SERIES. *Modular* means that each topic is presented in a separate book. Modularization produces compact, low-cost text and reference materials to suit any need and any course-organizational pattern. The flexibility afforded by modularization permits schools to select program components best suited to needs and time available.

Exploration means that the world of technology is opened up for investigation. The MET Series is not merely a group of "how-to" books. It does explain how to work with wood, fix engines, and make electronic devices. But exploration implies more than that. The MET Series opens to readers the creative spirit of men and women, the excitement of discovery, and the rewards of patient research. MET readers will study many and various technologies. At the same time, creative "hands-on" activities will reduce abstractions and increase motivation.

1952517

Preface

MANUFACTURING PROCESSES: CERAMICS provides a basic understanding of ceramics as an industrial material. You will learn about the many different types of ceramic materials necessary to produce the thousands of manufactured ceramic products. Raw materials, processing techniques, and finishing methods are discussed in detail. Many of the finished products are described and illustrated to show the wide and varied uses of ceramics in home and industry.

Contents

UNIT 1
CERAMIC MATERIALS — 1

UNIT 2
FORMING METHODS — 20

UNIT 3
DRYING AND FIRING — 45

UNIT 4
DECORATING AND FINISHING — 65

UNIT 5
REFRACTORIES AND GLASS — 82

UNIT 6
USES OF CERAMIC MATERIALS — 97

GLOSSARY — 114
INDEX — 117

UNIT 1

This unit acquaints you with the types and preparation of ceramic raw material.

1. You will be able to list the common raw materials used in ceramic products.
2. You will be able to explain how the raw materials are mined.
3. You will be able to explain how the raw materials are refined.
4. You will be able to describe the properties and characteristics common to ceramics.
5. You will be able to describe the three-point system used to make ceramic bodies.

WHAT ARE CERAMICS?

Ceramics are made from minerals that come from the earth's crust. These minerals are taken from the earth and crushed, refined, mixed, shaped, dried, and *fired* (heated to a high temperature) to form a ceramic product. The firing causes the ceramic to harden into a usable product.

Ceramic products are very common around your home and in many industries. Steel is made in furnaces lined with a ceramic product. Electronic parts made from ceramics are used in radios, televisions, transformers, and power transmission lines. Windows, dinnerware, plumbing fixtures, bricks, concrete, plaster, floor tiles, and wall tiles are only a few of the many ceramic products manufactured.

Origin and Development of Ceramic Material

The earth is made up of three main layers: the *crust, mantle,* and *core.* The surface, or *crust,* is approximately 10 to 40 miles (16 to 64 kilometers, or km) deep. Under the crust is a layer called the *mantle*. The mantle is made of a soft, hot rocklike material, approximately 1,800 miles (2,900 km) thick. The center part of the earth, the *core*, is approximately 4,300 miles (6,900 km) in diameter. Two parts form the core: a liquid shell and a solid center. Figure 1–1 shows the earth's layers.

The weight of the crust causes an extreme amount of pressure to develop in the mantle. Since the mantle is made of a soft material, the crust has a tendency to slide and shift over the mantle occasionally. This movement often causes cracks to develop in the crust. The soft material in the mantle is forced out of the cracks. Often when the crust develops a crack in the surface, the hot liquid is forced through, and settles on the earth's surface. Deposits of cooled and solidified liquid are called *lava*. Volcanoes are examples of the hot liquid being forced through the surface. Figure 1–2 pictures a volcano during the time when lava is being forced through the earth's surface. Sometimes the liquid is not forced through the surface, but remains in the crust. Then it is called *magma*.

As the lava and magma cool, they harden, forming *igneous rock*. Igneous means formed by heat. The weight and heat from the mantle, building up for thousands of years, cause the igneous rock below the surface to change. Extremely lengthy periods of heat and pressure form a new rock called *metamorphic* rock.

For millions of years, the igneous rock on or close to the surface is exposed to the weather. Rain, snow, ice, and gases from the atmosphere cause the rock to break and wear apart. Streams, rivers, and lakes wash the broken parts of the rocks away. Breaking, wearing, and washing away of the particles is called *erosion*. Erosion is a constant process, and thousands of years may pass before any change can be noticed by the naked eye. Figure 1–3 shows the process of rock development and erosion.

Formation of the rock, heat, pressure, and eroding are all necessary to produce the raw material for ceramics. *Clay*, which is a direct result of the erosion process, is the most important raw material used in ceramics. The two types of clay are *residual* and *sedimentary*. Residual clay deposits result from rock that was not washed away and decayed plant and animal life. Sedimentary clay deposits occur when the washing away of rock and plants has taken place.

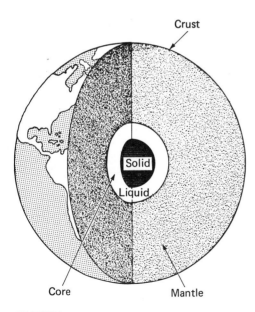

FIGURE 1–1
Layers of the earth.

Since earliest time, we have known that clay can be shaped and hardened into permanent shapes. All ancient civilizations have left vases, drinking pots, and bricks behind. Almost all of these products are made from clay.

Why Study Ceramics?

The ceramic industry is ranked among one of the largest in the United States. The industry employs hundreds of thousands of people. Yearly income in the form of wages and salaries adds millions of dollars to our economy. Products manufactured from ceramics are found in almost every building, factory, and vehicle in the world. Over 1,400 companies produce articles made of ceramic materials in the United States alone.

Classification of Ceramics

Ceramics are classified, or grouped, by four different methods. The first method classifies ceramics according to the processes used to produce a finished product. The second method groups ceramics by the characteristics or properties of the raw materials and the finished products. Thirdly, ceramics are classified by how the raw materials are formed. Finally, ceramics are classified by the resulting finished products. Table 1–1 shows a few examples of each type of classification system.

PROPERTIES OF CERAMICS

Properties of ceramics refer to the characteristics of the material. Characteristics are the qualities that separate

FIGURE 1–2
A volcano erupting. (Victor Ynglebert-DeWys, Inc.)

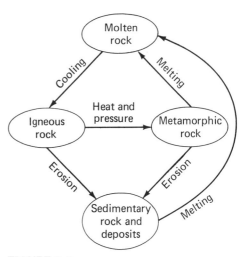

FIGURE 1–3
Rock development cycle.

TABLE 1-1 METHODS OF CLASSIFYING CERAMICS

Manufacturing Processes	Properties
Hand forming	Mechanical (strength)
Machine forming	Electrical
Casting	Thermal (heat)
Pressing	Miscellaneous

Products	Origin
Whitewares	Igneous
Refractories	Metamorphic
Glass	Residual
Abrasives	Sedimentary
Coatings	
Constructional	

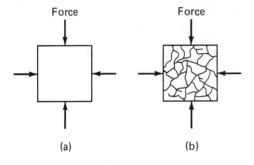

FIGURE 1-4
Compressive strength. The same force applied to A and B. B cracks because it has low compressive strength.

ceramics from other materials. Many types of ceramics are used in modern industry. Each is selected for a specific purpose, based on special characteristics. Properties of ceramics are mechanical, thermal, electrical, and miscellaneous. Each of these groups is discussed separately in the following sections.

Mechanical Properties

The *mechanical properties* of ceramics explain how the material reacts or is affected when exposed to various types of force. Compressive strength, tensile strength, impact strength, hardness, plasticity, and brittleness are mechanical properties.

Compressive Strength. *Compressive strength* is the ability of a material to resist a crushing force. If a weak crushing force is applied to a material and it breaks, the material is said to have a *low* compressive strength. However, if a great deal of force is needed to break the material, it has a *high* compressive strength. Figure 1-4 shows the difference between low and high compressive strength.

Tensile Strength. *Tensile strength* is the ability of a material to resist a pulling or stretching force. Material being tested is clamped between the jaws of two separate vises. The vises are mechanically forced in opposite directions until the material breaks. The force required to pull the material apart is a measure of the tensile strength. Force is measured in pounds per square inch (psi). Ceramics are very low in tensile strength. Figure 1-5 shows the type of force required to measure tensile strength. An industrial-type tensile tester is shown in Figure 1-6.

Hardness. The ability of a material to resist indentation, or penetration from another object is called *hardness*. If material A can scratch material B, then A is harder than B. Diamonds are the hardest known natural material. Other materials are compared to a diamond's hardness. A scale is used to rank materials from soft to hard. *Moh's scale* has been developed for this purpose. Table 1-2 shows a few common materials arranged according to hardness on Moh's scale. Number 1 on the scale is the softest, and Number 10 is the hardest.

Hardness is measured on many kinds and styles of hardness testers. Figure 1-7 shows a hardness tester that is sometimes used by industry. A pointed object is forced into the surface of the material being tested. The amount of indentation caused by the pointed object and the force used are measured. Hardness of the material is determined by using these measures. Hardness is directly proportional to the amount of indentation and the force used.

Impact Strength. When a material will not break easily when hit by a moving object, it has good *impact strength*. If you threw a hard ball at a piece of glass, the glass would probably break. However, if you threw the same ball at a brick wall, it would have little effect on the wall. The brick wall has greater impact strength than the glass. Ceramic products have a wide range of impact strengths. Some break easily. Others react like the brick wall. The range of impact strengths for ceramics is easily understood when you learn that both glass and brick are ceramic products. Impact strength is usually measured by a swinging pendulum or by dropping a ball on the material being tested.

Plasticity. Another mechanical property of ceramics is called *plasticity*. Having plasticity means that a material

FIGURE 1-5
Testing tensile strength.

FIGURE 1-6
Tensile tester. (W. J. Patton, *Materials in Industry*, 1968, p. 63. Reproduced by permission, Prentice-Hall, Inc., Englewood Cliffs, N.J.)

TABLE 1-2 MOH'S HARDNESS SCALE

Hardest	10	Diamond
	9	Corundum
	8	
	7	Quartz
	6	
	5	Glass
	4	
	3	
	2	Gypsum
Softest	1	Talc

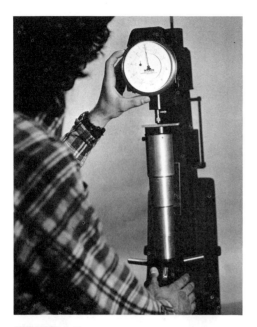

FIGURE 1-7
Rockwell hardness tester. (B. A. Tietz)

can be shaped without breaking. Care should be taken not to confuse *plastic material* with the product *plastic*. Molding clay is a plastic material because it can be reshaped many times. Most ceramic materials are moldable during some stage of manufacturing. When a ceramic material is fired to a high temperature, it becomes hard and no longer can be molded. The material cannot be reshaped without cracking after it has been fired. Fired ceramics are brittle materials. *Brittleness* is the opposite of plasticity. All ceramics are brittle after firing.

Thermal Properties

The *thermal properties* of materials refer to their characteristics during and after heating. *Thermal conductivity* and *thermal shock resistance* are the two major thermal properties of ceramics.

Thermal Conductivity. *Thermal conductivity* refers to the speed at which heat travels through the material. To test a ceramic for thermal conductivity, a sample is heated at one end, and the temperature at the other end is measured. Also, the time required for the temperature to increase at the measuring point is recorded. When the temperature increases in a short period of time, the material is a good *conductor*. However, if it takes a long time for the temperature to increase, the material is a good *insulator*. Ceramics are good insulators. As insulators, they are used widely in the construction of furnaces.

Thermal Shock Resistance. When a material can change temperature rapidly without cracking, it has good *thermal shock resistance*. Ceramics vary in thermal shock resistance. Some must be heated very slowly, or they will break. Other ceramics can be changed from hot to cold quickly without breaking. Low thermal shock resistance can be seen by placing an ice cube in a glass of hot water. The ice cube will begin to crack in a few seconds. Cracking occurs because the ice cube has low thermal shock resistance.

Electrical Properties

Electrical conductivity refers to the speed at which electricity can travel through a material. This property is similar to thermal conductivity. The electricity transferred, rather than the heat transferred, is measured. When the electricity is allowed to travel easily, the material has good electrical conductivity. When materials do not allow electricity to pass easily, they are good electrical insulators. Ceramics are good electrical insulators. Many electrical products have ceramic parts because of the insulating ability of ceramics.

Other electrical properties of ceramics include the amount of power loss by an insulator and the use of ceramics as capacitors.

Miscellaneous Properties

In addition to the mechanical, thermal, and electrical properties, ceramics have many other properties, which we will classify as *miscellaneous*.

Surface Characteristics. Ceramics have different *surface characteristics* which is the feel or touch of the product. Ceramics can be grouped into three types of feel qualities: (1) smooth, (2) rough, and (3) slippery. Products such as insulators, porcelain, and glass are smooth and feel slippery. Bricks, cement, and abrasives usually feel rough.

Color. Another miscellaneous property of ceramics is the *color* of the product. Some ceramic products can be pure white after processing. However, most of them contain other minerals which change the color. Color can also be changed by the following methods:

1. coating the product with a colorant before firing.
2. coating the product with a colorant after firing.
3. adding other minerals to the product.
4. heating the product to a specific temperature.

Common methods used to change or alter the color will be discussed later in the text.

Flow. *Flow* properties refer to the ease with which a liquid can be poured. Some liquids, like water, pour easily. Other liquids, like oil, pour much slower. Some ceramic products are made from liquid clay, called *slip*. The ease of pouring the slip is important. If the slip is not easy to pour, the ceramic product may be ruined.

Shrinkage. Many of the products made from ceramics contain water during some stage of manufacturing. After the product is shaped, it is allowed to dry. While drying, the shaped article becomes smaller. The reduction of size, due to water loss, is called *shrinkage*. Shrinkage is very important, since the size of the final product will be smaller. The amount of shrinkage depends on the amount of water in the clay, the size of the article, and the thickness of the article. Figure 1–8 shows a clay product while it is wet and after drying. Notice the difference in the size after the product is dry.

Porosity. *Porosity* refers to how close the ceramic particles are to each other. A *dense* material has ceramic particles very close together. When the product is dense, it will not absorb much water. A material is *porous* when the particles are farther apart. A porous product is lighter in weight than a dense one. The porous product will absorb water more readily than a dense product.

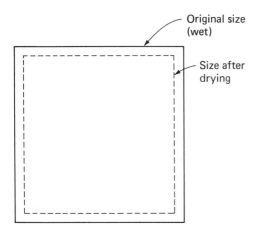

FIGURE 1-8
Size of ceramic tile before and after drying.

RAW MATERIALS

Clay

The most commonly used raw material in ceramic products is *clay*. However, some products such as glass and electronic parts use little, if any, clay. Three common classifications of clay are: kaolin, ball clay, and fireclay.

Kaolin. *Kaolin* is the purest type of clay available. It is white in color before and after processing. Kaolin was first used in China and is sometimes called China clay.

Kaolin is classified into two groups. The first group is a *residual clay* (found where the mineral was first deposited). The second type of kaolin is called *sedimentary*. Sedimentary kaolin consists of particles that were washed away from the original deposit of the mineral. As the clay particles were washed downstream, they were cleaned and settled in slow-moving lakes and lagoons.

Kaolin is found in Europe, China, and the United States. Most of the kaolin found in the United States comes from Florida, Georgia, and the Carolinas. Most common uses of kaolin are in the manufacturing of refractories and white clay products.

Ball Clay. *Ball clays* are sedimentary clays and are added to a ceramic product to increase the product's strength. Ball clays have a fine grain size. The clay particles are very small. After processing, the ball clays are cream-colored rather than white like kaolin. Ball clays are also highly plastic. When mixed with other clays, the mixture can be shaped and formed easily.

Ball clays are found in Kentucky and Tennessee. England is also a big producer of this type of clay. Major uses are in the manufacturing of electrical components and dishes.

Fireclay. The third type of clay is called *fireclay*. Fireclays are used to make firebrick and *refractories*. A refractory is a material that can withstand very high temperatures. The main use of ceramic refractories is for lining furnaces. Fireclays are dark in color, usually a reddish or reddish-brown.

Fireclays are divided into three groups. The first group contains *flint*. Flint is very hard and low in plasticity. The addition of flint to clay causes the ceramic material to be very hard and low in plasticity. Most flint clay is used to make brick. Flint clay is found in Missouri, Illinois, Tennessee, and Pennsylvania.

Plastic clays, the second group, are softer and more plastic than the flint clays. Plastic clays are used because of their ability to mix easily with water.

The last group of fireclays are called high *alumina clays.* This type of clay contains aluminum and oxygen. High alumina clays are used for firebrick and refractories. A shortage of high alumina clay exists in the United States, and most of them have to be imported.

Stoneware Clays. *Stoneware clays* are used for art pottery and ceramic products found in the chemical industry. Stoneware is generally buff (tan) in color. Occasionally it may be gray or brown. The color depends upon the amount of impurities present in the clays and the method used to fire the product.

Fluxes

Fluxes are mixed with clay and other minerals when making ceramic products. A flux is a mineral that melts at a low temperature. The purpose of using a flux is to bond, or hold together, the other raw materials. It can be compared to paste, or glue.

Feldspar. *Feldspar* is the most commonly used flux because of its abundance in the United States. Feldspar comes from an igneous rock usually found combined with quartz. Clay and feldspar are mixed, formed into the desired shape, and fired to a high temperature. The feldspar melts and forms a molten liquid, which causes the particles of clay to stick together. When the product cools, the feldspar hardens, giving strength to the product.

Feldspar is found in many different rocks. Granite is the best source of feldspar. In the United States, deposits of feldspar are located in the New England states and in North Carolina.

Cornish Stone. *Cornish stone,* also called China stone and purple stone, is still another flux. It is a mixture of China clay, feldspar, quartz, and other minor minerals. Most of it is found in Great Britain; however, a similar type mineral is found in the Carolinas.

Nepheline Syenite. *Nepheline syenite* can be used in place of feldspar. Nepheline syenite has an advantage over feldspar, which is a lower melting temperature. The main source of nepheline syenite is Canada.

Bone Ash. A flux made of animal bones is called *bone ash.* The bones are dried and crushed, prior to being used as a flux. Bone ash is mixed with clay to make a special type of ceramic material used in quality dinnerware. Bone ash is commonly found in England.

Other Raw Materials

In addition to the clay and fluxes described, other minerals are also used in ceramic products. Many minerals are added to improve the properties of products. Only the main materials used are discussed.

Silica. *Silica* is the second most plentiful element on earth. Oxygen is first. In its natural state, silica is found in quartz and sandstone. Flint is also con-

UNIT 1: CERAMIC MATERIALS

FIGURE 1-9
Gypsum rock. (Georgia-Pacific Company)

TABLE 1-3 ADDITIONAL MINERALS AND THEIR USES

Minerals	Uses
Alumina	Spark plugs
Chromite	Brick
Corundum	Abrasives
Diamond	Abrasives
Dolomite	Refractories
Graphite	Refractories
Lead	Glass
Magensite	Insulators
Sodium	Glass

sidered a form of silica. Silica is mixed with clay and fluxes for the following reasons:

1. to help maintain the shape of the product during the firing process.
2. to help reduce shrinkage of the product.
3. to prevent cracks from occurring in the product.

Silica is also the most important ingredient in the making of glass.

Talc. *Talc* comes from the steatite rock, found on the east coast of the United States. It is used in the manufacturing of wall tile. Talc is the softest material on Moh's hardness scale.

Lime. *Lime* comes from limestone, a very abundant mineral in the United States. After the mineral is refined, the lime is used to make plaster and mortars. A high-quality lime is also used to make glass.

Gypsum. Plaster of Paris, an important material for the ceramics industry, is used to make molds, building blocks, and wallboards. Plaster is made from *gypsum,* another ceramic raw material. Figure 1-9 shows gypsum rock before refinement.

Table 1-3 lists some of the other minteals and their principle uses in the ceramics industry.

MINING AND REFINEMENT

Raw materials used to produce the various ceramic products are found in rock formations formed by the rock cycle. Since raw materials are not distributed evenly in the earth's crust, we must first locate large deposits of them. Searching for mineral deposits is called *prospecting.*

Prospecting requires a knowledge of *geology* (the study of the earth, its formation, and changes). Knowledge in this science helps scientists determine where certain minerals may be found. With the use of instruments and photographs of the earth's surface, the scientist looks for evidence of the required materials. When an area which may yield the mineral is identified, holes are drilled in the earth's surface. A sample of the soil and rock is examined, and the various minerals are identified. When the desired raw material is present, a series of holes are drilled in the surrounding area. These enable the prospectors and scientists to determine if an ample supply of the needed material is present. Prospectors drilling test holes are shown in Figure 1–10.

Types of Mines

After the mineral deposits have been located in the quantities and qualities needed, some method must be used to remove them. Since some of the deposits are close to the earth's surface, and some are deeper in the crust, different methods are used.

Open-pit mines are used when the raw materials are close to the surface. It is necessary to remove only the top layer of soil, called *overburden,* to expose the needed materials. Bulldozers, scrapers, and power shovels are used for this purpose. The raw material itself may be removed by power shovels or scrapers. Sometimes, however, the raw materials are very hard. Then, explosives are used to break apart the hard mineral deposits. After removal from the earth, the minerals are transported to the processing plant. Figure 1–11 shows a side view of an open-pit mine. A picture of an open-pit mine is shown in Figure 1–12.

To remove a layer of the material deposit found deep in the ground, an *underground mine* is used. Usually this type of mine is called a *shaft mine.* A shaft or tunnel is dug toward the deposit until the raw material is reached. The mineral is then broken apart by drilling or blasting. Raw materials are brought to the surface in small railroad cars. A side view of a shaft mine is

FIGURE 1–10
Drilling test holes. (Thomas J. Barbre of Frederic Lewis)

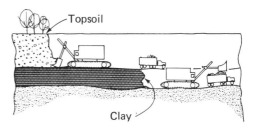

FIGURE 1–11
Cross-section of an open-pit mine.

FIGURE 1–12
An open-pit mine. (Rotkin, P.F.I.)

shown in Figure 1–13. Figure 1–14 pictures a miner preparing to drill a hole in gypsum. An explosive will be inserted in the hole, causing the rock to break apart by blasting.

Another method of removing the raw material, usually clay, is by *hydraulic mining*. Hydraulic mining uses a stream of high-pressure water to wash the clay down a side of a hill. The water and clay mixture is collected at the base of the hill. This mixture is then pumped to the surface for further processing.

Crushing and Grinding

After the raw materials are mined, they have to be crushed or ground into smaller particles. Crushing and grinding larger particles into smaller, usable particles is called *comminution*. Several methods of comminution are employed in preparing the raw materials for use in ceramic products.

One method uses force. Raw materials are passed between two large rollers that exert great amounts of force to break the hard materials. Force can also be applied to two flat plates with the raw material between them. Figure 1–15 shows an example of rollers being used to crush the raw material.

Impact-type force can also be used for crushing the raw materials. Machines for this method have a series of hammers which strike the raw material. Impact from the hammers cause the raw material to break apart. Lumps of hard clay can be broken up in the school laboratory with hammers. Safety glasses must be worn whenever you are working in the school laboratory.

Abrading, or the use of a rough surface to reduce the size of raw materials, is another method of communition. A rough file can be used to abrade material in the school laboratory. When the file is drawn across the material, small particles are removed. Figure 1–16 illustrates how abrading occurs.

Grinding, still another means of communition, is accomplished in a ball mill or industrial grinder. The *ball mill* consists of a hollow cylinder, filled with the raw material and hard steel balls. The cylinder is rotated, causing the balls to tumble over one another, grinding the raw material between them. A ball mill is shown in Figure 1–17. A picture of an industrial-type grinder is shown in Figure 1–18. The center section revolves over the raw material, causing it to be crushed and ground into smaller particles. Such processes are widely used in industry.

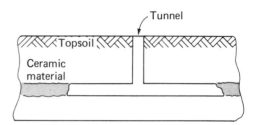

FIGURE 1–13
Cross-section of a shaft mine.

FIGURE 1–14
Miner preparing to drill a hole. (Georgia-Pacific Company)

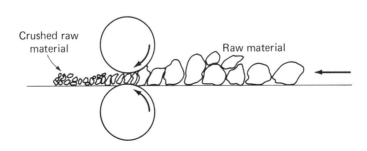

FIGURE 1–15
Roll crusher.

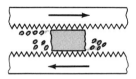

FIGURE 1-16
Abrading.

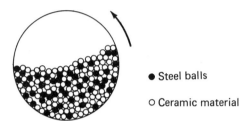

FIGURE 1-17
A ball mill.

FIGURE 1-18
An industrial grinder. (J. C. Steele and Sons, Inc.)

Size Classification

After the particles of raw material are reduced in size, they are classified according to size for further processing. Some processes require that the material be crushed into a fine powder. Particles are separated by sifting them through a screen. Screens are made up of a number of specific sized holes. The number and size of the holes in the screen determine the *mesh size* (number of openings per linear inch). Sizes of the screens range from 2 to 400 mesh, or have from 2 to 400 openings per linear inch. Large numbers have more openings per linear inch; however, these openings are very small. In large-numbered screens, only small particles can be sifted. Small-numbered screens allow larger particles to pass. Separation into particle size is made possible through the use of this process. Figure 1-19 illustrates the difference between a number 10 mesh and a number 30 mesh screen size. Different mesh sizes and the size of particles that can pass through are listed in Table 1-4. Small screens can be purchased for use in the school laboratory from various suppliers. A flour sifter can be substituted to demonstrate the principle.

Refinement

Screened materials often contain impurities that may ruin the ceramic product. Removal of the impurities is essential to maintain high standards. This is called *refinement*. Common methods of refinement are *magnetic separation* and *froth flotation*.

Magnetic separation employs a magnet to remove iron or magnetic materials from the crushed raw materials. Powdered materials are passed over magnets which remove iron particles. Magnetic separation can only be used when the material contains iron or magnetic particles. Figure 1–20 shows how the magnetic-separation process works.

When refining is done by *froth flotation,* crushed materials are mixed with water and some other liquid that form bubbles. The bubbles cause lightweight particles to float to the top, and heavier ones to sink to the bottom. By this separation technique, unwanted materials can be discarded. The froth-flotation process is shown in Figure 1–21.

Raw materials can also be treated with chemicals to remove impurities. This process is much more expensive than the others. Because of cost, this process is seldom used.

Refined materials are allowed to dry before packaging. A typical dryer is shown in Figure 1–22. Drying is accomplished by loading a tilted, rotary furnace with raw materials. Heat is passed over the wet material while the dryer is being rotated. As the dryer rotates, the material slides down the cylinder. It dries as it slides to the opposite end, where it is removed.

CERAMIC BODIES

After the raw materials are refined, they are ready for further processing. Ceramic products are used for various purposes and need to meet specific requirements. Some raw materials are used to make bricks, others are used for high-quality dinnerware. As stated previously, raw materials are com-

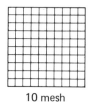
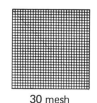

10 mesh   30 mesh

FIGURE 1–19
Difference between 10-mesh and 30-mesh screens.

TABLE 1-4 MESH SIZES FOR SCREENS

Mesh Size	Particle Size	
3	.260 inches	6.6 millimeters
10	.065	1.6
20	.032	0.8
60	.009	0.2
100	.005	0.14
200	.002	0.07
400	.001	0.03

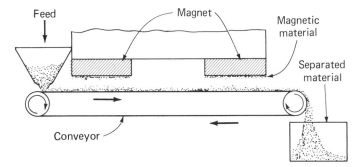

FIGURE 1–20
Operation of a magnetic separator.

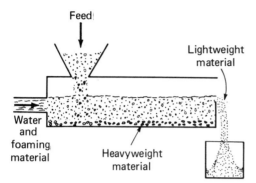

FIGURE 1-21
Froth-flotation process.

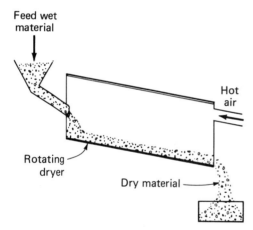

FIGURE 1-22
A rotary dryer.

bined in various proportions to improve properties, workability, and color. The mixture of the raw materials is called a *ceramic body*.

Many clays found in this country can be used just as they come from the ground. However, the properties and workability of the bodies are improved by the addition of other raw materials.

The following ingredients are commonly used in ceramic bodies:

1. Clay is used as the plastic material.
2. Flint and quartz are added to reduce shrinkage.
3. Fluxes are used to bond the clay particles.

Triaxial Bodies

Clay, flint, and feldspar are mixed in various proportions to form ceramic bodies for specific purposes. A method known as the *triaxial-body system* is used to show the proportions of each ingredient. Figure 1-23 shows a triangle used in the triaxial system. Each point on the triangle represents 100 per cent of the raw material listed at that point. The triangular plot shows the relationship, or the quantity, for each of the three ingredients used to make a triaxial body.

Ceramic bodies that are based on the triaxial system are shown in Figure 1-24. Each type of body is located close to the position that indicates its composition.

Other Ceramic Bodies

The triaxial system is used for clay products; however, other products require different types of bodies. Glass, refractories, coatings, abrasives, and cement are different types of ceramic bodies.

Glass has a different structure than most ceramics. However, both natural and synthetic glass are considered ceramic products. Certain characteristics make glass a desirable material for in-

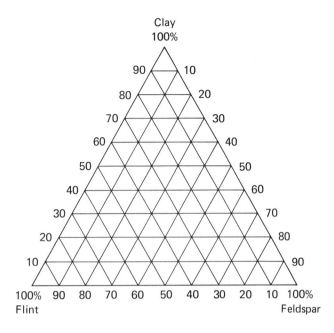

FIGURE 1-23
Triangle used in triaxial system.

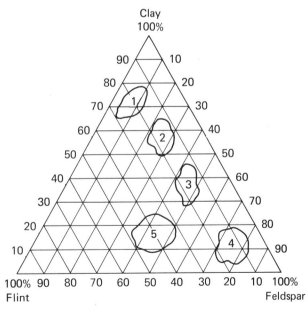

FIGURE 1-24
Triaxial system.

1. Wall tile   3. Electrical porcelain   5. Hard porcelain
2. Floor tile 4. Dental use

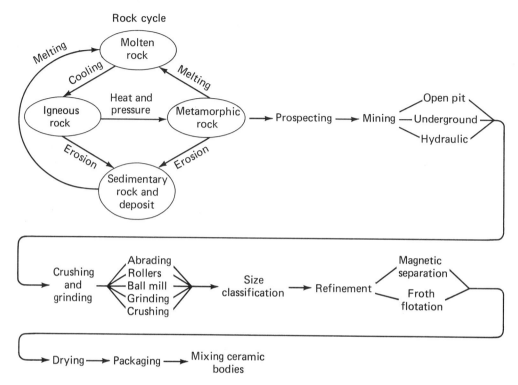

FIGURE 1-25
Summary of preparing ceramic material.

dustrial products. A prime characteristic of glass is its *transparency* (clearness). This makes it an excellent material for glasses and windows. The materials required to make glass are different than the triaxial system. These bodies are called *glass bodies.*

Refractories are ceramic products that can withstand extreme heat without failure. In order to maintain their characteristics at high temperatures, special bodies are developed.

Coatings require a different body composition. Porcelain enamel is the most common coating used on metals. Protection and appearance are the most desirable characteristics of ceramic coatings.

Abrasives used in grinding wheels and abrasive papers for cutting and polishing require different body compositions. Certain raw materials, such as flint and emery, make very good abrasives. Newer synthetic bodies produce abrasives of superior quality.

Cement and plaster are ceramic bodies that do not require firing to make them hard. Their prime characteristic is durability, making them excellent in building construction.

There are numerous ceramic bodies used to make finished products.

More detail concerning each of the bodies follows later.

As a summary, Figure 1–25 illustrates the processes used to produce ceramic bodies from naturally occuring minerals. Later units discuss how the ceramic bodies are formed, decorated, and used.

ACTIVITIES

1. Build Your Vocabulary:
    a. ceramics
    b. igneous
    c. plasticity
    d. porosity
    e. geology
    f. comminution
    g. ceramic body
    h. triaxial bodies
2. Perform the various methods of comminution, in the school laboratory, using a hammer, file, rolling pin, and lumps of hard dry clay. (If a fine mesh screen is available, sift the small clay particles to obtain particles of uniform size.)
3. Mix clay, flint, and feldspar (powdered form) to total 100 grams. Try different proportions of each, recording the amounts of each. Place the mixtures in separate containers, and save them for later use.
4. Identify and list various ceramic products in your home, and try to identify the main ceramic raw material used.

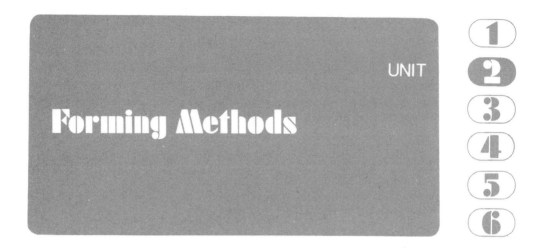

UNIT 2: Forming Methods

This unit acquaints you with the shaping and forming of ceramic products.

1. You will be able to classify ceramic products according to how they are formed.
2. You will be able to explain how clay is prepared for the various forming methods.
3. You will be able to describe the common forming methods used to produce a ceramic product.

Relationship to Other Materials

Various methods are used to form ceramic products. Many are similar to the processes used to produce plastic and metal products. Ceramic products can be shaped by hand, machine, or a combination of hand and machine methods. Various forming processes include casting, bending, pressing, rolling, and extruding. Vacuum and blow forming, used to produce plastic products, are also used for some ceramic products. Although the forming processes are similar, there are differences in the types of equipment and forms of the raw materials used.

Classification of Forming Methods

Before ceramic raw materials can be formed, they are usually mixed with water. The amount of water mixed with the raw materials determine which forming process is used. *Water content* is the per cent of water added to the raw materials.

Manufacturing processes used to form ceramic products are classified by the water content of the ceramic body. When very little water is used (a low water content), greater force is required to shape the product. As the water content increases, the pressure needed to form the product decreases.

Forming methods are classified by the water content, as: (1) liquid; (2) soft plastic; (3) stiff plastic; (4) semidry; and (5) dry. Table 2-1 shows the approximate water content for each type of clay body.

Table 2-2 summarizes the major forming methods used for each type of clay body. It should be noted that some processes are used for more than one type of clay body. Processes used for forming glass and some other products will be discussed in later units.

CLAY PREPARATION

Clay and ceramic bodies can be purchased in liquid, plastic, and dry forms. When slip or plastic clay forms are purchased, they are ready for use. Dry forms have to be weighed, mixed, and processed before use. Many interesting experiments can be performed with dry forms.

Manufacturers of ceramic products usually buy dry forms. They can mix their own materials and form a ceramic body having the desired properties. Recall from Unit 1 that a ceramic body is a mixture of the raw materials. In this unit, the term *body* will also be used to identify the form of the body.

Preparation of the ceramic body is the first step in forming a ceramic product. Bodies have to be prepared in various proportions of raw material to water. Weighing the raw materials is the first step in the preparation of the body.

Weighing and Mixing Ingredients

Accuracy in measuring the raw materials is very important. Accurate scales are necessary to be sure the proper amounts of each materials are added. A double-beam balance can be used. A balance is shown in Figure 2-1. For industrial use, a larger scale is needed. Accuracy is still very important in the larger scales.

Scales and balances usually use the *metric system* of measurement. You

TABLE 2-1 CLASSIFICATION OF CLAY BODIES BY WATER CONTENT

Clay Body	Water Content (in percentages)
Liquid (slip)	20–30
Soft plastic	15–25
Stiff plastic	5–15
Semidry	3–7
Dry	0–5

TABLE 2-2 CLASSIFICATION OF FORMING METHODS

Type of Clay Body	Forming Methods
Liquid	Solid casting
	Drain casting
Plastic (soft/stiff)	Hand molding
	Pinch
	Slab
	Coil
	Hand and machine throwing
	Machine molding
	Jiggering
	Jolleying
	Roller
	Extrusion
	Pressing
Semidry and dry	Pressing
	Hot pressing
	Isostatic forming

FIGURE 2-1
A balance used to weigh dry ingredients. (Carl Glassman)

TABLE 2-3 ENGLISH AND METRIC EQUIVALENTS

English		Metric
1 ounce	=	28.4 grams
1 pound	=	454 grams
2.2 pounds	=	1 kilogram
1 pint	=	.47 liters
1 quart	=	.95 liters

probably used the metric system in a mathematics or science class. Table 2-3 provides a short summary of the major English and metric equivalents.

Remember that a ceramic body can be made of many different raw materials. A sample body used in producing bath tubs and sinks may include:

- 31 per cent   Flint
- 26 per cent   Feldspar
- 24 per cent   Kaolin
- 19 per cent   Ball clay

Raw materials are weighed according to various body proportions. The weighed materials are then added to water. Before the raw materials and water are mixed, they are allowed to *slake*. Slaking is the period of time needed for the raw materials to absorb water. The amount of water varies with the mixture of raw materials. Usually 4 to 6 ounces of water are added for each pound of dry materials.

Raw materials and water are mixed after the slaking period. Mixing the clay and water is called *blunging*. Blunging is done with a machine that is similar to a milk shake mixer. Figure 2-2 shows a small blunging machine. For small batches, a household electric mixer can be used. Mixing is continued until all the materials are completely mixed. Proper mixing of the ingredients may take many hours. Lumps of dry ingredients or inadequate mixing may ruin a product.

From this step on, the processes vary for producing the different forms of clay bodies. Probably the most common method is to make *slip* (liquid clay). Other bodies are made by removing the proper amounts of water from the slip.

Liquid (Slip)

The mixture of clay and water must be filtered after mixing. A screen, having a fine mesh, is used to remove any lumps that may be present. Impurities or lumps in the slip will ruin or cause defects in the product. Manufacturers pass the slip over magnetic separators to remove iron-bearing impurities. The operation of the magnetic separator was described in Unit 1.

The water content of slip and soft plastic bodies is almost the same. In order to make the slip pour easily, a chemical is added. Chemicals added for this purpose are called *deflocculants.* When deflocculants are added to the slip, less water is needed. The advantages of using a deflocculant, rather than more water, are listed below.

1. Less water is used, therefore, the product will dry in less time.
2. Small clay particles will not settle to the bottom of the liquid.
3. Less water reduces the amount of shrinkage in the product.
4. The possibility of cracks in the final product is reduced. The more a product shrinks while drying, the higher the chance of cracks forming.

A number of deflocculants may be added to the slip. Common deflocculants are sodium silicate, soda ash, and sodium tannate. When deflocculants are added, the slip must be mixed again in the blunger. Figure 2-3 illustrates the steps necessary to make slip.

FIGURE 2-2
An industrial-type blunger. (Wedgwood)

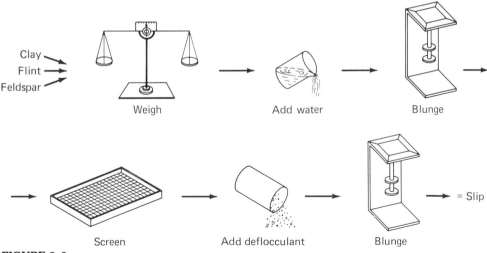

FIGURE 2-3
Process for making slip.

FIGURE 2-4
A filter press. (Wedgwood)

Plastic Bodies

Plastic bodies can be made by adding the proper amount of water to the dry ingredients, and then blunging. The clay is hard to mix because it is not a liquid. A common method used to make plastic bodies is to start with slip. The water is then removed to form a plastic body. Deflocculants are not added to the slip when making a plastic body.

Slip has a slightly greater water content than the plastic clay. In order to make plastic clay, some of the water must be removed. Filter presses are used to remove the excess water from the slip. The slip is poured into the filter press. The press is closed, and the water is forced out of the slip. When the process is finished, the press is opened and plastic clay is removed. Figure 2-4 shows a filter press.

In the school laboratory, a filter press may not be available. Another method to remove water is to use plaster bowls or containers. Plaster absorbs the water from the slip. The slip is poured into the plaster container. Water is removed by the plaster, forming plastic clay. The clay is then ready for further use. Figure 2-5 shows the process of making a plastic clay body.

Dry and Semidry Bodies

Making a semidry body is similar to making a plastic body. The clay is dried for a longer period of time to reduce the moisture content of the clay. When making a dry body, a different method is used. Raw materials and water are blunged, as described for a slip. Again, deflocculants are not added to this mixture. The slip is then sprayed into a heated chamber. The heat and compressed air dries the clay. The dry clay body is screened to obtain particles of the same size. After screening, the body is ready for further processing. Figure 2-6 illustrates the process used to produce a dry body.

Consistency of Clay Bodies

Water content is the major factor in classifying clay bodies for the various forming processes. The term *consistency* describes how the clay body looks, feels, and how easily it can be shaped. Table 2-4 describes the consistency of the various body forms.

Additional information about clay

UNIT 2: FORMING METHODS 25

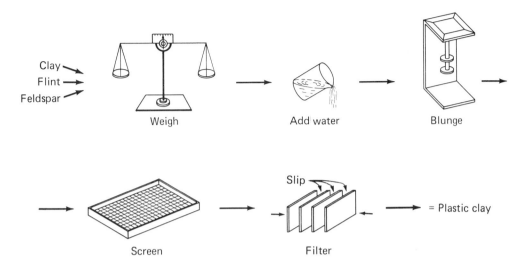

FIGURE 2-5
Process for making plastic clay.

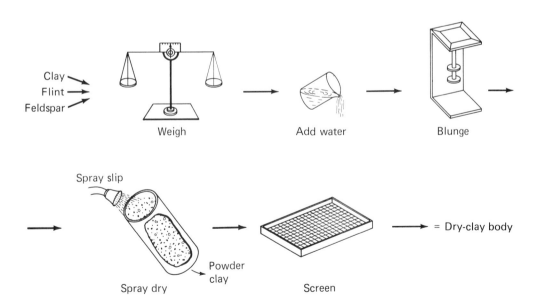

FIGURE 2-6
Process for making dry-clay bodies.

UNIT 2: FORMING METHODS

TABLE 2-4 CONSISTENCY OF CLAY BODIES

Type of Clay Body	Consistency
Liquid	Liquid; looks like thin cream; must have water removed during the forming process; will shrink a great deal after forming.
Soft plastic	Easy to form; will stick to your hands; will stick to itself with very little pressure; water needs to be removed while shaping; product will sag or deform unless the water content is reduced; will shrink a great deal after forming.
Plastic	Will not stick to your hands; will stick to itself with little pressure; can be shaped with hand pressure; will not sag when shaped; will shrink slightly.
Stiff plastic	Needs more than hand pressure to shape it; will not sag; very little shrinkage.
Semidry and dry	Needs high pressure to be shaped; very little, if any, shrinkage after shaping.

consistency and workability is presented with each forming method. After the product is formed, changes take place in the clay because of water loss. Changes in the body after forming will be presented in Unit 3.

CASTING

Casting is a forming process that uses liquid clay called *slip*. Casting is a common industrial process. Slip is poured into a plaster mold. Water is absorbed by the plaster, leaving a clay product inside the mold.

There are two major types of ceramic castings. In *solid* castings, the slip is poured into the mold and left until it becomes solid. In *drain* castings, the slip is poured into the mold for a period of time. Then the slip is drained from the mold. Drain casting is used to produce a hollow product that has a thin wall. The type of product needed determines which casting process is used.

Molds used in the ceramic-casting process are made from plaster. Plaster is used because it easily absorbs water from the slip. Molds can be made of one or more pieces. If more than one piece is used, the parts must be held together when it is used. String, rubber bands, and straps may be used to hold the parts of the mold together. It is important to align the mold parts before use. If they are not aligned, the product will probably be ruined.

Solid Casting

Making a solid product requires the solid-casting process. After the mold is aligned and fastened together, the slip is poured into the mold. Water is removed from the slip by the plaster. As the water is removed, the level of the slip drops. If the level drops too low, the casting will have a hole in it. To ensure that enough slip remains in the mold, an extra amount is added. Parts of the mold has an area which holds extra slip. This area is called the *spare.* Figure 2–7 shows a cross-section view of the mold. Figure 2–8 illustrates a solid casting in the mold.

Drain Casting

A hollow product having a thin wall is made in a drain-casting mold. Slip is poured slowly into the mold. While the slip is in the mold, the plaster draws out the water, forming a shell next to the mold. When the proper wall thickness is reached, the mold is drained of the excess slip. After a short period of time, the clay shell inside the mold becomes hard enough to remove. Figure 2–9 illustrates the drain-casting process.

Casting slip and molds can be purchased from various suppliers. Molds and the slip can also be prepared in the laboratory. Casting is a major industrial process that can be performed on a small scale in the school laboratory. The following list gives instructions for making a drain casting.

1. *Preparation.* Check the molds to be sure that they are clean. Place the parts of the mold together. Be sure that they are aligned properly. Secure the parts of the mold using rubber bands. Mix the slip and

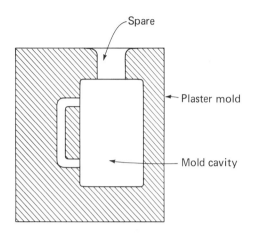

FIGURE 2–7
Cross-section of a mold.

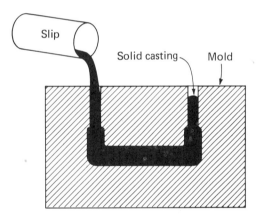

FIGURE 2–8
Solid casting.

FIGURE 2–9
Drain casting process.

28 UNIT 2: FORMING METHODS

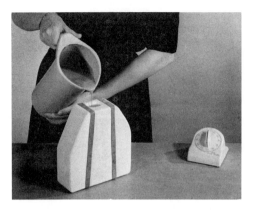

FIGURE 2-10
Slip being poured into a mold. (Duncan Ceramic Products, Inc., P.O. Box 7827, Fresno, Calif. 93727)

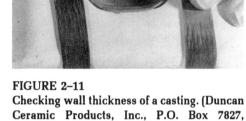

FIGURE 2-11
Checking wall thickness of a casting. (Duncan Ceramic Products, Inc., P.O. Box 7827, Fresno, Calif. 93727)

screen it to remove any lumps or impurities.

2. *Casting*. Pour the slip in a slow, steady stream into the mold. When the slip is poured too slowly, the casting will have lines on the surface. If the slip is poured too fast, it may splash on the inside of the mold. Splashing causes small defects on the surface of the casting. Figure 2-10 shows the slip being poured into the mold.

   Keep the mold full of slip, until the desired wall thickness is reached. Small products should have a wall about 1/8-inch thick. A thicker wall is desired for larger products. Time needed for slip to remain in the mold depends on: (a) the size of the product (more time is needed for large objects); (b) dryness of the mold (less time is required when the mold is dry); and (c) the water content of the slip, (more water in the slip requires a longer casting time).

3. *Draining*. When the desired wall thickness is reached the excess slip may be removed. Turn the mold over slowly, and drain the excess slip into a container. Be sure to drain the mold slowly. If the slip is drained too fast, the wall of the casting may pull away from the mold and be damaged. Allow the mold to remain upside down (at a slight angle) to remove all of the slip. Figure 2-11 shows how to check wall thickness.

4. *Removal*. Before the product can be removed, it is allowed to set until the slip becomes *leather-hard.* It is leather-hard when enough moisture has been removed from the clay that the product can be handled without deforming. Excess clay around the mold opening should be removed. Figure 2-12 shows how to remove the extra slip with a knife. The rubber bands can now be removed. Open the mold slowly and carefully. Be sure to lift

FIGURE 2-12
Removing excess slip from the mold. (Duncan Ceramic Products, Inc., P.O. Box 7827, Fresno, Calif. 93727)

the top half of the mold straight up. Figure 2-13 shows how to open the mold. Carefully lift the product out of the mold. Remember that the casting is soft and must be handled carefully.

PLASTIC SHAPING

Hand-forming Methods

Plastic shaping consists of hand, machine, and a combination of hand- and machine-forming processes. Ceramic articles made by ancient civilizations were formed by hand methods. You are probably familiar with hand methods. Remember shaping small objects with modeling clay in elementary school? The processes described in this section are very similar to forming modeling clay. Instead of using an oil-base clay, we are using a water-base clay.

A medium plastic clay is used for

FIGURE 2-13
Opening the mold. (Wedgwood)

FIGURE 2-14
A wedging board. (American Art Clay Co., Inc.)

hand molding. Before an article can be molded, the clay must be wedged. Wedging makes the clay uniform by removing air pockets. Wedging is also used to dry the clay if it is too wet. Figure 2-14 shows a wedging board.

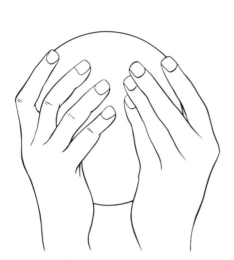

FIGURE 2–15
Roll clay into a ball.

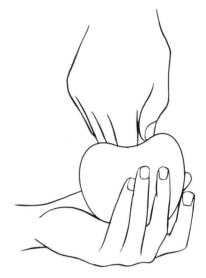

FIGURE 2–16
Starting the object.

The board contains a heavy plaster base, a backing board, and a piece of fine wire. Wedging boards can be purchased or constructed in the laboratory. When constructing a wedging board, be sure that it is solidly built. The board will take a lot of rough use while wedging.

Hold the lump of clay to be wedged in your hands. Pass the clay over the wire, and cut the lump in half. Throw one half of the lump on the plaster base. Throw the other half on top of the first. Pick up this lump and cut it again. Repeat this procedure 15 to 20 times. Check the clay after wedging, by cutting it again. If there are no air bubbles and the clay looks the same across the cut, it is ready for use. If the clay becomes too dry while being wedged, sprinkle water on it and continue wedging. Wet clay can be dried by laying it on the plaster base of the wedging board.

Pinch Forming. *Pinch forming* is the simplest type of hand-forming method. While the process is easy, its industrial use is very limited. The process is slow, and it is very difficult to produce a number of objects exactly the same using this process. The process used to pinch-form a small bowl is described in the following list.

1. Roll a lump of plastic clay into a ball that is 2 or 3 inches in diameter.
2. Hold the ball in one hand. Press the thumb of your other hand into the center of the ball.
3. Rotate the ball, shaping it into a small bowl.
4. Sponge the bowl with a damp

UNIT 2: FORMING METHODS 31

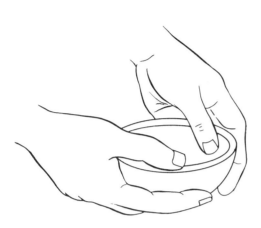

FIGURE 2-17
Shaping the object.

sponge. Sponging makes the surface of the object smooth.

Figures 2-15 to 2-17 illustrate the pinch-forming process.

Slab Forming. The use of clay slab to construct a ceramic article is called *slab forming*. As an industrial process, slab forming is seldom used. However, it does provide the student with the opportunity to construct objects using a different forming method. After the clay is wedged, it is ready for making the clay slab. The process used to construct a small box made from slabs is described as follows.

1. Develop a set of plans (working drawings).
2. Place the clay between two strips of wood on a work table. Use a rolling pin to roll the clay. The wood strips allow the clay to be rolled to an even thickness. (A piece of cloth may be placed below and above the clay to keep the clay from sticking to the table and rolling pin.)
4. Remove the cloth from the clay. Draw lines on the clay according to the plans. Use a knife and cut the clay along the lines.
5. Remove the slabs and set them in position for assembly. If the clay is too wet, allow the slab to dry enough to be handled without deforming.
6. Roughen the edges that will touch each other. Apply slip with a brush to the areas that were roughened. Attach the slabs by using the slip like a glue. Sticking plastic clay together with slip is called *slip welding*.

For objects that have thick walls, *grog* may be added. Grog is a fired clay that has been crushed into small particles. Since the grog has been fired, it will not shrink. The grog is added to unfired clay to reduce the shrinkage of the object formed. Add the grog before the clay is wedged. Figure 2-18 illustrates the slab-forming process.

Coil Forming. Hand forming of ceramic articles can also be done with a process called *coil building*. The process is simple and has very little use to industrial producers. The steps in coil building are listed below.

1. Roll a lump of clay into a coil about 3/8 to 1/2 inch in diameter. Care should be taken to make the coil the same diameter along its length.

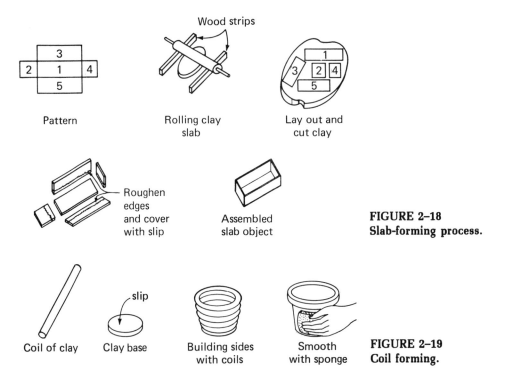

FIGURE 2-18
Slab-forming process.

FIGURE 2-19
Coil forming.

2. Place a small slab of clay on the table, and cut the slab to the desired shape for a base.
3. Roughen the outer half inch of the base. Shape the coil to fit the base. Attach the coil to the base by slip welding. Continue to make coils and place them on top of each other.
4. Make smaller coils, about 1/4 inch in diameter. Place the small coils between the larger ones, on the outer surface of the object.

Figure 2-19 illustrates how to make a coil-formed article.

The preceding hand-forming methods show how to make a few of the many articles by hand. Industrial use of hand-forming methods is rare because of the time involved. Also, making two objects exactly the same is very difficult using hand methods.

Hand and Machine Forming

A *potter* is an individual who produces ceramic articles by hand on a machine. The machine is called a *potter's wheel* or just *wheel*. Making ceramic articles on the wheel requires a great deal of skill. An experienced potter can make two or more articles per minute. Many years of practice are required to produce objects of good quality on the wheel. Making ceramic articles on the potter's wheel is called *throwing*. Throwing is used to make a round, hollow object. The object is shaped by hand while it revolves.

Various types of wheels are available from many suppliers. Some wheels are powered by the potter, others are powered by a motor. Foot-operated wheels are called *kick* wheels. The operation of the wheel is the same for both kick and motor-driven wheels.

Plastic clay is centered on the throwing head. The throwing head revolves, allowing the potter to form the desired object. Because of the water in the clay, the wheel and its parts must be made of rust-proof material. Aluminum is the most common material used in making a wheel. Figure 2–20 shows a kick wheel, and Figure 2–21 shows a motor-driven wheel.

Plastic clay is wedged and centered on the throwing head. The inside bottom of the object is formed first. The potter then forms the side walls, by applying hand pressure to the revolving clay. Finally the article is shaped to the desired design. Figures 2–22 to 2–27 show the potter throwing an object. Figure 2–28 illustrates what the cross-sectional shape of an object looks like while being thrown.

Throwing is a process used by many artists to produce hollow clay articles. The throwing process is also used by industry to produce specialized products. Industrial use of throwing is very limited, because of the skill and time required to throw an object.

Machine Molding

Clay Preparation. In the industrial setting, the clay is prepared by power equipment. Because greater amounts of clay are needed, larger capacity machines are used. The dry ingredients are weighed and placed in a motor-driven high-production mixer.

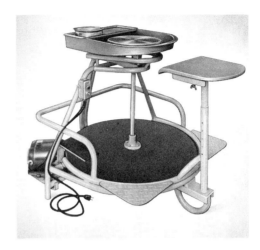

FIGURE 2–20
Kick-type potter's wheel. (American Art Clay Co., Inc.)

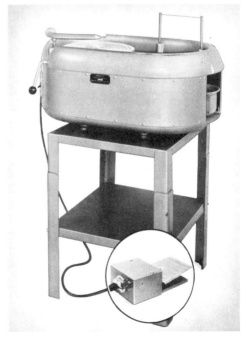

FIGURE 2–21
A motor-powered potter's wheel. (American Art Clay Co., Inc.)

34 UNIT 2: FORMING METHODS

FIGURE 2-22
Preparing clay for throwing.

FIGURE 2-23
Centering the lump of clay.

FIGURE 2-24
Starting to form the wall.

UNIT 2: FORMING METHODS 35

FIGURE 2–25
Building the wall.

FIGURE 2–26
Forming the lip.

FIGURE 2–27
Finishing the surface.

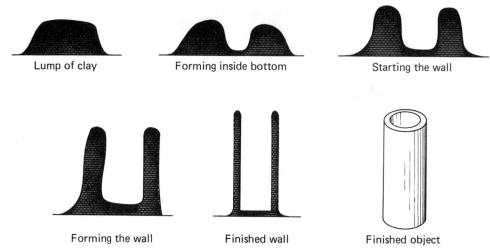

FIGURE 2-28
Stages in forming a thrown object.

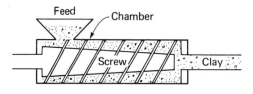

FIGURE 2-29
Cross-section of a pug mill.

When the raw materials are mixed with water, the clay is wedged by hand. Hand wedging is a slow proccess. It produces small amounts of prepared clay. Rather than hand wedging, a *pug mill* is used to produce larger amounts of usable plastic clay. A pug mill is a machine that mixes and removes air from the clay body. The pug mill has a feed hopper, chamber, screw shaft, and discharge end. Clay is placed in an open chamber on the machine. The chamber has a shaft with a series of knives, resembling a screw. The shaft revolves and cuts the clay. Because of the screw design, the clay is forced through the chamber to an opening at the end. Figure 2-29 shows a cross-section of the pug mill. Figure 2-30 shows an industrial-size pug mill.

If the clay is placed in a pug mill having a vacuum, the clay becomes more plastic. A vacuum pug mill is used to mix the clay and remove all the air from the clay body.

Clay that dries out during processing can be reused. As the clay dries, it hardens and needs to be chopped into smaller particles. Figure 2-31 shows a waste chopper. The clay is placed in the machine and is chopped by a series of revolving knives. After the clay is chopped, it can be placed in the pug mill with water and remixed.

Jiggering and Jolleying. Mechanical processes used to produce a number of identical articles are called *jiggering* and *jolleying*. Jiggering is the process used to shape the *inside* of a product. Jolleying is the process used to shape the *outside* of the product. Figure 2-32 illustrates the jiggering method, and Figure 2-33 the jolleying process. They are used to produce most of the ceramic dinnerware manufactured.

The process is similar to throwing. Clay revolves on a wheel head like the potter's wheel. Instead of a potter shaping the product by hand, a *template* is used to form the product. A template is a silhouette of the product being shaped.

Jiggering uses prepared clay from the pug mill. The refined clay body is weighed and cut to the proper size. Before the clay is shaped on the jiggering machine, it is flattened. Flattened clay looks like a pancake and is called a *bat.* It is very important that the bat is free of air bubbles. The operator places the bat on the throwing head. The machine is started, causing the bat to rotate. Then an arm is lowered, by the operator, that contains the template. Templates are made of wood, plastic, or metal. Water is sprayed over the clay to

FIGURE 2-30
A pug mill. (J. C. Steele and Sons, Inc.)

FIGURE 2-31
A waste chopper. (J. C. Steele and Sons, Inc.)

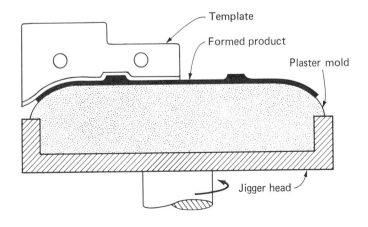

FIGURE 2-32
Jiggering.

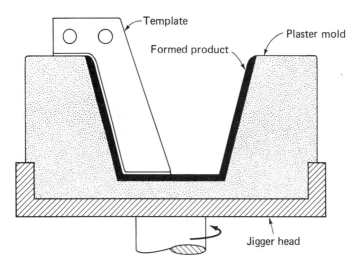

FIGURE 2-33
Jolleying.

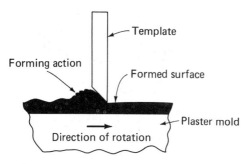

FIGURE 2-34
Forming a jiggered product.

lubricate it. While the clay is revolving the template scrapes the clay, causing it to be forced against the mold, then allowed to dry. The process is then repeated to produce another identical product. Figure 2-34 shows how the clay is scraped between the template and the mold.

Jolleying is similar to jiggering. However, the mold is shaped differently. The clay from the pug mill is not flattened into a bat. Instead, a lump is placed in the mold and pushed against the mold wall by the operator. The machine is started, and the template is lowered. Again, the template causes the clay to take the shape of the mold and template. The number of products produced by jiggering and jolleying is dependent upon the size of the product and the experience of the operator. Figure 2-35 shows the jolleying process.

For industrial production, automatic jiggering and jolleying machines are used. Figure 2-36 shows an automatic jiggering unit. A heated plate is lowered, pressing the clay on the mold. The template is then lowered, and it finishes shaping the product. Production machines having more than one forming head can produce up to 1,400 products an hour.

Jiggering and jolleying are industrial processes that can be done in the school. Small machines are available from many suppliers. Figure 2-37 shows a small jiggering unit that can be used in the school laboratory.

FIGURE 2-35
Jolleying. The operator trims the top of shaped ware. (Lenox)

FIGURE 2-36
An automatic jolleying machine. (Wedgwood)

Roller Forming. A roller machine is also used to form dinnerware. *Rolling* differs from jiggering, in that a template is not used. Clay is placed on the mold and is shaped by a roller that forces the clay into the desired shape. A bat is not needed, therefore, the time used to produce the bat is saved. Roller machines are either semiautomatic or fully automatic.

Extrusion. *Extrusion* is a machine-forming process used to produce a continuous product having the same cross-sectional shape. The product is called the *extrudate.* Extrudates are cut to the proper length and allowed to dry. Figure 2-38 illustrates various cross-

FIGURE 2-37
A small jiggering machine for school use. (American Art Clay Co., Inc.)

sectional shapes that can be extruded.

Two major types of extruders are *auger* and *piston.* An *auger* extruder has a large screw in an enclosed cham-

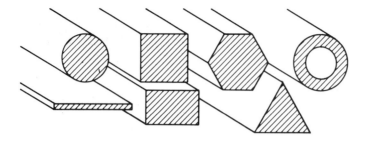

FIGURE 2-38
Shapes that can be extruded.

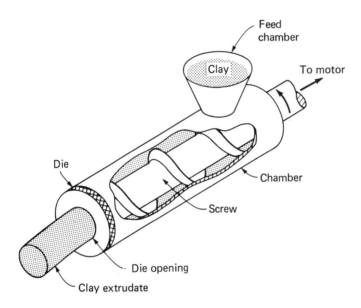

FIGURE 2-39
Operation of an auger extruder.

FIGURE 2-40
An industrial-type extruder. (J. C. Steele and Sons, Inc.)

ber. The screw revolves, and forces the clay to the end of the chamber. At the end of the chamber is a *die.* A die contains a hole of the shape desired for the extrudate. Clay is forced through the die and takes the shape of the hole. The extrudate is cut by machines to the proper length. An auger extruder is the same as a pug mill. Figure 2–39 shows the operation of an auger extruder. Figure 2–40 shows an industrial extruder that is used to make ceramic pellets.

Piston-type extruders make up the second group. Figure 2–41 shows the piston extruder and its operation. Plastic clay from the pug mill or mixer is loaded into the extruder's chamber. A piston compresses the clay, forcing it through the die opening. The piston moves back, and the chamber is reloaded with clay. Even though the operation uses a back-and-forth movement of the piston, the extrudate is a continuous product. A cake decorator or cookie maker can be used to demonstrate the operation of a piston-type extruder.

The extrusion process is used to make bricks, drainpipes, insulators, spark plug shells, roof and floor tiles. In addition, extruded shapes are also used to feed prepared clay bodies to other machines used to form ceramic products. Industrial-size extruders are capable of producing ceramic pipe up to 30 inches in diameter.

As mentioned, the extrudate is continuous and must be cut to the proper length. Figure 2–42 shows an in-

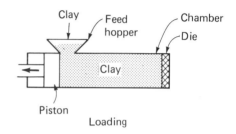
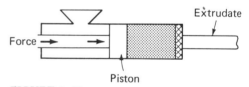

FIGURE 2–41
A piston extruder.

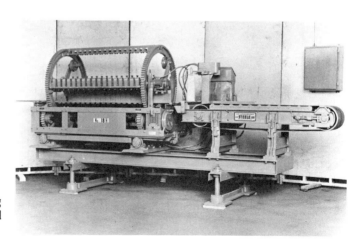

FIGURE 2–42
An automatic brick-cutting machine. (J. C. Steele and Sons, Inc.)

dustrial brick-cutting machine. Cutters are capable of cutting up to 15,000 bricks per hour.

STIFF-PLASTIC FORMING

In stiff-plastic forming, clay with an average of 13 per cent water is used. Because of the low moisture content, a higher forming pressure is required. When high pressures are used, the stiff-clay body will flow easily into the desired shape.

Stiff-plastic forming can be done on an extruder. Both auger-and piston-type extruders are used. The same type of products made by extruding plastic clay are made with the stiff-plastic clay bodies.

Bricks can be produced on a press-type brick-production machine. Vertical pug mills mix the clay body and deposite the clay in a series of molds. Molds are then positioned below the press. The clay is pressed into the molds, forming bricks. The shaped bricks are removed from the molds and taken to a second press. The purpose of the second press is to make the bricks an accurate shape and add *surface texture*. Surface texture on bricks is the design pressed into the brick.

Because a low-moisture-content clay is used, the products can be stacked on top of each other. Products made by stiff-forming methods dry quickly. Therefore, the complete process to make finished products will take less time and money.

SEMIDRY AND DRY FORMING

Semidry-forming methods requires that the clay body have 5 to 15 per cent water content. Dry-forming methods require that the water content of the body be less than 5 per cent. As previously mentioned, when low-moisture-content clays are used, higher forming pressures are needed. Pressures needed to form the body range from 10,000 to 100,000 pounds per square inch (psi). The dryer the clay body, the higher the pressure needed.

Pressing is the common forming method. Rather than plaster molds

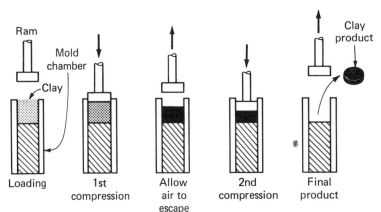

FIGURE 2-43
Operation of presses.

used in plastic forming, steel molds are needed for semidry and dry forming. Steel molds are used because of the high forming pressures needed.

Low water content in the bodies may cause problems during the forming process. The clay particles do not stick together easily. Therefore, an ingredient called a *binder* is added. Dry ceramic powder may stick to the mold walls. *Antistick* ingredients are added to prevent the powder from sticking to the molds.

Pressing is the most common forming method used on low-water-content bodies. In order to reach the high pressures needed, the presses are operated by *hydraulic* or *pneumatic* power. Hydraulic presses use force from a liquid under pressure. Pneumatic presses are operated by compressed air.

Operation of the presses is illustrated in Figure 2–43. The clay body is loaded into the press. Pressure is applied to the mold, compressing the clay. The mold is opened to allow air to escape. After the air escapes, the press is closed again to compress the body. When the product is formed, it is removed from the press. Presses for semidry and dry forming can be either semiautomatic or fully automatic.

Hot Pressing. *Hot pressing* is similar to dry pressing. High pressure plus heated molds are used. The molds are heated from 500 to 1200°C (932 to 2192°F). The body is pressed into the desired shape. Heat from the mold causes the product to become hard and dense. When products are formed by heat and pressure, creating a product that does not have to be fired, it is called

FIGURE 2–44
Donst 33-ton compacting press. (AC Compacting Presses)

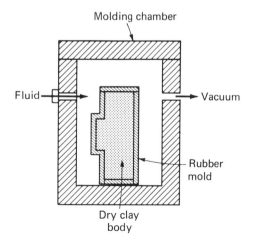

FIGURE 2–45
Isostatic pressing.

sintering. Hot pressing produces ceramic products used in cutting tools and abrasives. See Figure 2–44.

Isostatic Pressing. *Isostatic pressing* is a relatively new type of forming method. Liquid under high pressure is used to shape the product. Both the rubber mold and clay body are placed in a chamber. Air is pumped from the chamber. Then a liquid is forced into the chamber, under high pressure, forcing the rubber mold to compress the clay. Spark plug shells are the most common product produced by isostatic-pressing methods. Figure 2–45 illustrates the isostatic-pressing method.

ACTIVITIES

1. Build Your Vocabulary:
    a. water content
    b. slake
    c. blunging
    d. deflocculant
    c. throwing
    f. jiggering
    g. extrusion
    h. isostatic pressing
2. Using the dry ceramic bodies prepared in Unit 1 (activity number 3), prepare plastic ceramic bodies. Use the plastic bodies to form objects using the pinch-, slab-, and coil-forming methods.
3. Construct a wedging board.
4. Develop a series of charts that show the different forming methods.
5. Construct small models that can be used to demonstrate the various forming methods.

Drying and Firing

UNIT 3

This unit acquaints you with the drying and firing of shaped ceramic products.

1. You will be able to describe various methods used to dry ceramic products.
2. You will be able to describe the steps of the firing cycle.
3. You will be able to dry and fire ceramic ware properly.
4. You will be able to describe methods used to measure the temperature of a kiln.
5. You will be able to describe the differences between the various kinds of kilns.

Water is an important ingredient in most ceramic bodies. Solid particles of the ceramic body are surrounded by the water. Water is used to cushion and lubricate the solid particles. A water cushion around the particles allows the body to be shaped easily. Recall that, when a high water content is used, the body is formed with very little effort. When the water content is reduced, greater pressures are needed to shape the body.

After the ceramic body is shaped, the water is no longer needed. Formed ceramic products (ware) have to be fired to very high temperatures. At high temperatures, the ceramic particles harden and stick together. If water is present in the body during firing, the object may explode. The water must be removed before firing. The drying step removes the excess water.

DRYING

Drying is a simple but important step in producing a finished ceramic product. As the water is removed from the shaped ware, the object will shrink. Rapid drying can cause uneven shrinkage. Uneven shrinkage occurs when one part of the object shrinks more than the other parts. When an object shrinks unevenly, it will deform or crack. Slow, even drying is recommended.

Potters and artists usually set shaped ware aside to let them dry slowly. Sometimes their objects are covered with plastic wrap or a wet cloth to slow the drying period. Slow drying is impractical for large ceramic manufacturers. They are concerned with the speed and quanity of production. To speed up the drying process, automatic dryers are used. The automatic-drying process is carefully controlled to prevent cracking the ware.

Stages in Drying

Solid ceramic particles are surrounded by water. As the water *evaporates,* the ceramic object dries. *Evaporation* is a process in which air absorbs the moisture in the form of a vapor. Thus, evaporation increases the moisture in the air, while reducing the per cent of water in the ceramic ware. Figure 3–1 illustrates the various stages of drying. Ceramic particles are separated by the water cushion. As the object starts to dry, water evaporates from the surface. During evaporation, the water cushion is removed, allowing the particles to move closer together. When the solid particles move closer together, the object shrinks. *Shrinking* of the object means that it becomes smaller.

As the solid particles on the surface touch, water is trapped in between the solid particles below the surface. Since the water is not exposed to the surface, it dries more slowly A series of

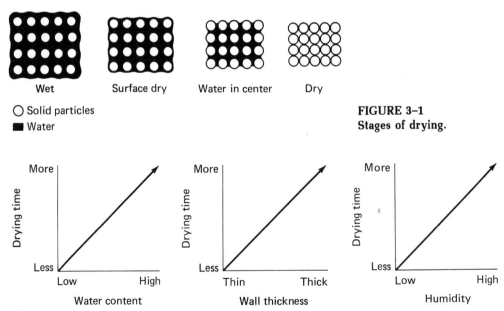

FIGURE 3–1
Stages of drying.

FIGURE 3–2
Factors affecting drying time.

small pores form in the object. Pores allow the remaining water to move toward the surface. This movement is a very slow process.

A number of factors affect the speed of drying. The first factor is the original *water content* of the products. High-water-content bodies, such as slip or soft plastic, require longer drying periods. The wall thickness of the object is the second factor. Products with thick walls require longer drying periods. A third factor in drying is *humidity*. Humidity is a measure of the water content in the air. High humidity means that the air contains a lot of moisture. Longer drying periods are needed when the air has a high humidity. Figure 3–2 shows the factors affecting drying time.

Warping

Formed ceramic ware can deform or warp after being shaped. Various factors are responsible for the object warping. Warping occurs when shaped ware is exposed to a force. Some of the reasons that cause a ceramic object to warp are listed below.

1. Improper wedging before the ware is shaped. Raw materials are mixed during wedging. If the materials are not mixed completely, stress may result. The stresses caused by incomplete mixing will cause the ware to warp.
2. Uneven wall thickness of the formed ware. When the ware is formed, the wall may be thicker at some places. A stress is developed because of the difference in thickness.

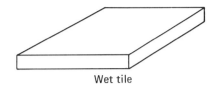

FIGURE 3–3
Unevenly dried tile.

3. After the object is formed, it must be handled carefully. Rough handling can cause stresses in the ware.
4. Drying can cause the object to warp. Ware must dry slowly and evenly. Uneven drying will cause sections of the ware to shrink before other sections. As sections shrink, they tend to pull the other sections out of shape. Figure 3–3 illustrates how a ceramic tile will warp if it dries unevenly.

Cracks

Cracks form in the ceramic ware for the same reasons that cause warping. However, stronger stresses are present when the ware cracks. When an object dries unevenly, it has a tendency to crack. Figure 3–4 illustrates how a cup can crack around its rim. Cracking around the edge can be caused by the edge drying before the base. A bowl with a cracked base is illustrated in Figure 3–5. The base cracked because

FIGURE 3–4
Crack in the rim of a cup.

FIGURE 3–5
Bowl with a crack in the base.

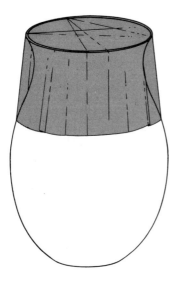

FIGURE 3–6
A vase covered with plastic to slow drying.

the rim dried first. When the rim shrunk, it forced the base to deform and crack.

Cracks can be prevented by drying the object evenly. To eliminate cracks in the rim or edge, the object should be covered. Plastic wrap or a wet cloth is used to cover the edge. Areas that dry faster than other sections should be covered to slow the drying. Plastic wrap or a wet cloth slows the drying period. Figure 3–6 shows the use of plastic wrap or cloth to slow the drying. A cover, like a box, can also be used. The cover is placed over the object. When using this method, be sure that the box is supported at the base. Air must be allowed to enter the box. Figure 3–7 shows how to cover and support the box.

When the base of an object dries slowly, another method can be used. Place the ware on blocks of wood. Air can then pass below the base, aiding in its drying. The object can also be placed on a screen, which allows air to circulate below the screen. Both methods can be used to speed the drying of the base. Figure 3–8 shows how to support ware on blocks of wood.

Types of Dryers

Commercial ceramic products are formed on automatic machines. This produces a large number of shaped ware in a short period of time. Allowing the objects to air-dry takes too much time. Space and time would be wasted for days, weeks, or even months while the ware dries. To speed the drying, automatic dryers are used.

Methods of Heating. Industrial-size dryers use heat and controlled humidity to dry ceramic ware. Heat is supplied by one of three methods. *Conduction, convection,* and *radiation* are the three sources of heat.

Conduction dryers use heat from heated floors or shelves in the dryer. Hollow tubes are built into the floor or shelf. Steam is forced into the tubes and heats the surface. Ceramic ware is placed on the heated surface. Heat passes through, or is conducted through, the ceramic ware. Water is forced out of the ware by the heat. Fans are used in the dryer to circulate and dry the air. Conduction dryers have a disadvantage. Ceramic objects are placed on a heated surface which may cause uneven drying. To prevent cracking and warping, very low temperatures are used. Because of the low heat used, the drying is slow. Conduction dryers are used mainly for building bricks and other special shapes.

Convection dryers use heated air to dry the ware. Moisture is removed from the ware by evaporation. Drying by convection is done in a number of steps, which are listed below.

1. Heated humid air is forced into the dryer. The air is used to heat the ware to an even temperature. During this step the ware will not dry because of the humid air.
2. The air is dried slowly while it circulates in the dryer.
3. The air is then reduced in temperature. Cool dry air in the dryer will absorb the evaporated water. Heated ware tends to force water to its surface. Ceramic ware will dry evenly in a convection dryer.

Radiation dryers use reflected heat. Radiation is a new method of dry-

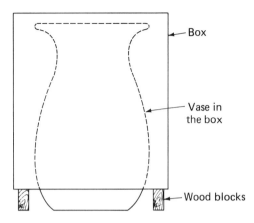

FIGURE 3-7
A vase enclosed in a box to slow drying.

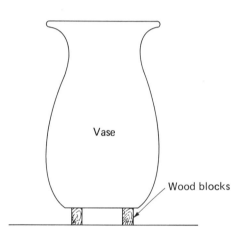

FIGURE 3-8
A vase supported on blocks, allowing the base to dry.

ing, which uses heat from light bulbs or a fire. Heat is reflected to the objects by mirrors or a highly polished surface. The heat reaches the ware and causes the water to evaporate. When heat is applied to the surface, the water in the center of the ware may not be removed. Radiation drying is used

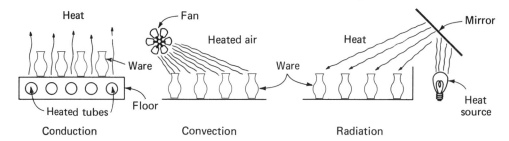

FIGURE 3-9
Methods of drying.

mainly on small or thin-walled objects. Figure 3-9 illustrates the principle of drying by conduction, convection, and radiation.

Batch and Continuous Dryers. Dryers are divided into two types. *Batch* dryers are the first group. A batch dryer is loaded with a group of ceramic objects. The entire group, or batch, is dried together and unloaded. When one batch is dried, another batch is loaded and dried. *Continuous* dryers are the second type. In a continuous dryer, the ceramic ware enters at one end of the dryer. The ware then moves through the dryer to the other end. During their movement, the ware is dried. This type of machine is constantly being loaded, drying, and being unloaded.

Batch dryers can be heated by conduction, convection, and radiation. Convection is the most common heating method used in batch dryers. Figure 3-10 illustrates the drying cycle in a batch dryer. Ceramic ware is loaded into a convection batch dryer. Warm humid air is forced into the dryer, heating the ware. The air is then dried and cooled. Heat in the ware forces water to the surface where it will evaporate. The cool dry air absorbs the evaporated water. After the entire drying process is complete, the ware is removed from the dryer. Batch dryers are made in various sizes to handle different-sized objects and loads.

Continuous dryers are made in various shapes and sizes. Each type of dryer is designed for certain ceramic products. Continuous dryers are constantly being loaded and unloaded. One type of continous dryer is called a *tunnel dryer.* The shaped ware is loaded on small flat railcars. Cars are on tracks similar to a railroad flat car. Once loaded, the cars move through a tunnel. As they move slowly through the tunnel, the ware dries. When the ware reaches the end of the tunnel, it is unloaded and ready for further processing.

There are two types of tunnel dryers. The first type has a series of rooms where the air contains different amounts of heat and humidity. The heating and drying process is the same as convection drying. The second type of tunnel dryer is shown in Figure 3-11. Ceramic ware enters the tunnel on railcars. Heated air is blown into the tunnel at the exit end. As the air passes through the tunnel, it becomes cool and humid. Ware entering the tunnel will not dry rapidly in the cool damp air. Speed of movement through the tunnel depends on the size and water content of the ware.

FIRING

Purpose

Up to this point, we have been concerned with shaping and drying the ceramic ware. By this time, most of the water has been removed from the ware during drying. The next step in producing a finished product is very important. To make the ware usable, the ware is heated to a high temperature. Heating to a high temperature hardens the ceramic material. In addition to becoming hard, the ware becomes water resistant and may change color. When the ware hardens in the fire, we say it *vitrifies.*

Without the firing process, the ceramic product would be useless. Before firing, the ware can be crushed, mixed with water, and reshaped. After firing, the ware cannot be reused except as grog.

The firing process is a critical step. A rapid increase in temperature will cause improperly dried ware to explode. Too high a temperature will cause the ceramic material to melt and deform. Too low temperature will not vitrify the ware properly.

Firing is done in a *kiln.* Kilns are similar to an oven. However, kilns can withstand very high temperatures. Kilns can be heated to 1260°C (2300°F).

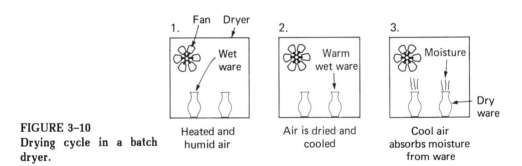

FIGURE 3-10
Drying cycle in a batch dryer.

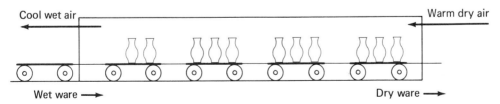

FIGURE 3-11
Continuous tunnel dryer.

There are many sizes and types of kilns available. No matter what type of kiln is used, the procedure for firing is the same.

Firing Cycle

When ceramic ware is fired, certain changes take place in the ceramic body. The process of firing the ware to the proper temperature is called the *firing cycle.* All of the ceramic bodies discussed so far go through the same stages of the cycle. The temperatures needed in each stage of the cycle depend on the type of ceramic body being fired. The major changes and steps of the firing cycle are listed below.

1. Removal of any water remaining in the ware after drying. Temperatures for this stage range from 100°C (212°F) to approximately 550°C (1022°F).
2. Organic (animal and vegetable waste) impurities in the ceramic body burn out. Temperatures from 400°C (752°F) to about 1000°C (1832°F) are needed to burn out the impurities.
3. Ceramic bodies mature from 900°C (1652°F) to 1400°C (2552°F). Maturing is the hardening and changing of the various properties in the body. Each type of body will mature or vitrify at different temperatures.
4. Cooling of the ceramic ware is the last step in the cycle.

Water Removal. Even after the ceramic ware has been dried, a small amount of water is still present. Various raw materials contain a small amount of moisture. Dryers do not reach the high temperatures needed to remove all of the water. The first stage of the firing cycle removes the remaining moisture.

Heating of the kiln should be slow at the start of the cycle. If the temperature increases too fast, the moisture turns to steam. Increases in steam cause pressure to build up in the ceramic material. The increased pressure from the steam may cause the ware to explode in the kiln. Steam also needs to be released from the kiln. Leaving the door of the kiln open a small amount allows the steam to escape from the kiln. Temperatures up to 550°C (1022°F) are usually required to remove all of the moisture. When this temperature is reached, the door of the kiln should be closed to speed up heating.

Removal of Organic Matter. When ceramic raw materials were formed by erosion, many impurities became part of the material. Some of these impurities are made of *decomposed* (rotten) animal and vegetable life. Refinement processes used to prepare the raw materials cannot remove these impurities. Impurities of decomposed materials are called *organic impurities.*

As the temperature increases in the kiln, the organic matter begins to burn out of the ware. Impurities begin to burn out at about 400°C (752°F). When the organic matter burns out, various gases are produced. Gases produced include carbon dioxide, carbon monoxide, oxygen, and hydrogen. Proper ventilation will prevent dangerous gases from making the workers ill. When the organic matter burns, producing gases, it is called *oxidation.* Different impurities oxidize at different temperatures. Most impurities oxidize between 400°C (752°F) and 1000°C (1832°F).

Maturing. During the final stage of the firing cycle, the ceramic body *matures*. In the previous stages, the water and impurities were removed from the ware. As the temperature increases, the ceramic material changes chemically. Clay particles become harder as the temperature increases. Fluxes in the body begin to melt. When the flux melts, it flows between the solid clay particles. A thin layer of liquid flux coats each particle in the ware. Other materials in the body also change to improve the properties of the ceramic ware. When all of the changes take place, we say the body has *matured.*

Ceramic bodies mature at various temperatures. Some mature at 1000°C (1832°F), others, such as porcelain, mature at 1300°C (2372°F). *Porcelain* is a very hard and dense ceramic product. Increasing the temperature needed to mature the body will cause additional changes in the ware. When the temperature is increased, the body becomes *denser.* A *dense* body is a body in which the ceramic particles are close together. Denseness also increases the strength of the ware. However, denser ware will shrink more. If the temperature is too high above the maturing temperature, the ware will *sag* or deform. If the temperature remains high or is increased, the product will melt and be ruined.

Cooling. After the kiln is fired to the proper temperature, it is shut off. Temperatures inside the kiln will drop slowly. It is important that the ceramic ware and kiln cool slowly. When cooling is too fast, the product may crack. During the cooling period, the flux hardens and holds the ceramic particles together. Cooling takes many hours because of the high heat used in firing. The ware should not be removed from the kiln before it reaches room temperature.

KILNS

A *kiln* is a special type of oven used to fire ceramic ware. High temperatures are needed to mature the ceramic body. Therefore, kilns have to be able to reach, and withstand, extreme temperatures. The kiln is lined with a firebrick called a refractory. Refractories are able to withstand the high heat in the kiln. Several types and sizes of kilns are produced for the different types of ceramic products.

Early Kilns

Firing ceramic ware dates back thousands of years. Ancient civilizations probably learned by accident that clay hardens when heated to high temperatures. Early firing methods were very primitive. Ceramic ware was placed in a small pit dug in the ground. Dried grass and twigs were used as the fuel. Fuel was placed in the pit below and around the ware. The fuel was ignited and allowed to burn. Heat from the fire caused the ware to mature. Figure 3–12 illustrates a simple pit used to fire early ceramic ware.

As time passed, various methods were developed to fire ceramic ware. Tunnels were dug in the ground and

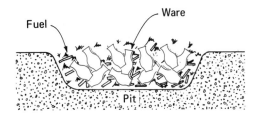

FIGURE 3-12
Early firing method.

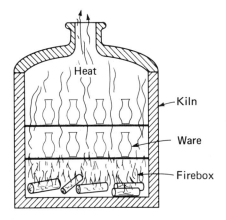

FIGURE 3-13
An updraft kiln.

used as firing chambers. Kilns were built above the ground and were used to fire the ware. Knowledge of *draft* made better use of the heat produced. Draft is a method used to circulate the heated air in the kiln. Heated air will rise above the cool air. By using draft, the heat produced by the fire can be used more efficiently.

TYPES OF KILNS

Updraft Kilns

The simplest type of kiln is an *updraft kiln.* Fuel is placed in a firebox. Ceramic ware is placed over the firebox in a firing chamber. The fuel is ignited and allowed to burn. Burning fuel in the box heats the air in the chamber. As the air moves up through the chamber, it heats the ware. Draft caused by the rising air pulls cool air into the firebox. Cool air is heated and rises through the chamber. The hot air is allowed to escape from the kiln through an opening in the top of the kiln. An updraft kiln is not very efficient. A great deal of heat is lost through the top of the kiln. Figure 3-13 illustrates a simple type of updraft kiln.

Downdraft Kiln

Downdraft kilns are more efficient than updraft kilns. Heated air is deflected through the ceramic ware. By using a downdraft kiln, the maximum use is made of the heated air. Figure 3-14 illustrates the principle of a downdraft kiln. Ceramic ware is stacked in the firing chamber. Heated air rises from the fire and is then reflected down through the ware. The heat is drawn through holes in the floor of the kiln and allowed to escape through a chimney.

Muffle Kiln

Ceramic products that are decorated with glaze need special care when being fired. In both the up draft and downdraft kilns, flames may touch the decorated ware. Glazed products and special ceramic products must not be touched by the flames. To protect the ware, a sealed firing chamber is used. The sealed chamber is surrounded by another chamber called a *muffle.* Hot air circulates through the muffle and

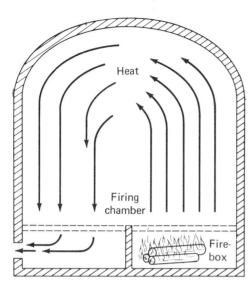

FIGURE 3-14
A downdraft kiln.

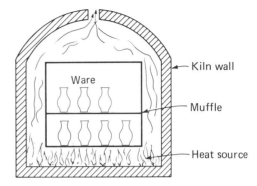

FIGURE 3-15
A muffle kiln.

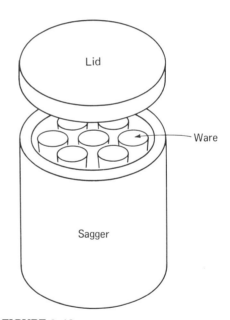

FIGURE 3-16
A sagger.

heats the firing chamber. Figure 3-15 illustrates the muffle kiln. Extra heat is needed to provide enough heat in the firing chamber. This type of kiln is not as efficient as the other types of kilns.

If decorated ware is fired in an updraft or downdraft kiln, the ware must be placed in *saggers*. Saggers are fired ceramic containers. The ceramic ware is placed in the sagger and covered during firing. Saggers prevent the flames from touching the decorated ware. Figure 3-16 illustrates the use of saggers.

Periodic and Continuous Kilns

Kilns used for production purposes are classified as either *periodic* or *continuous*. Periodic kilns are loaded, heated, cooled, and unloaded. The process is then repeated for each load. Continuous kilns are similar to continuous dryers. However, higher temperatures are used in the kiln. In a continuous kiln, the temperature can be kept constant in different sections of the kiln. Ware is loaded on small cars and moved through the kiln. During their movement, the ware goes through the firing cycle.

FIGURE 3-17
A front-loading kiln. (American Art Clay Co., Inc.)

FIGURE 3-18
A top-loading kiln. (American Art Clay Co., Inc.)

Periodic Kilns. A periodic kiln is the simplest and most commonly used kiln by schools, artists, and some industries. Various manufactures produce small periodic kilns. Figure 3-17 pictures a small front-loading kiln. Figure 3-18 pictures a top-loading kiln. Both kilns are capable of reaching 1260°C (2300°F).

Larger periodic kilns are used for production purposes. An *elevator kiln* is shown in Figure 3-19. Ceramic ware is placed below the kiln which is raised. The kiln is then lowered like an elevator, enclosing the ceramic ware. Once the top is lowered, the firing cycle begins. After the kiln is shut off and cooled, the top is raised. The kiln can be reloaded and fired again. It may take as long as 60 hours to complete the firing and cooling cycle.

Another type of periodic kiln is called a *shuttle kiln.* Instead of being raised like the elevator kiln, it is always at floor level. A door is provided to load and unload the kiln. Ceramic ware is moved into the kiln, the door is closed, and the ware is fired. When the ware cools, it is removed and the kiln is reloaded. Figure 3-20 shows workers moving ware into the shuttle kiln.

Continuous Kilns. There are a number of different designs of continuous kilns. Continuous kilns can be circular, oval, rectangular, or of a tunnel design. Figure 3-21 shows a tunnel kiln.

Ceramic ware is loaded on a flat car that moves through the kiln. At first, the air in the kiln is warm, but not hot. As the ware moves through the kiln, it reaches hot air. The ware moves through the entire firing cycle. By the time the ware reaches the exit end, it is cool. Figure 3-22 illustrates the operation of a tunnel kiln.

UNIT 3: DRYING AND FIRING 57

FIGURE 3–19
An elevator kiln. (Bickley Furnaces, Inc.)

FIGURE 3–20
A shuttle kiln. (Bickley Furnaces, Inc.)

FIGURE 3–21
A tunnel kiln. (Bickley Furnaces, Inc.)

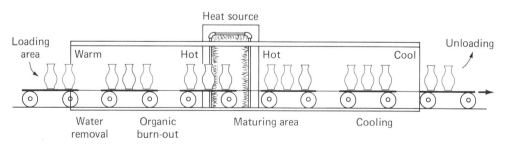

FIGURE 3–22
Operation of a tunnel kiln.

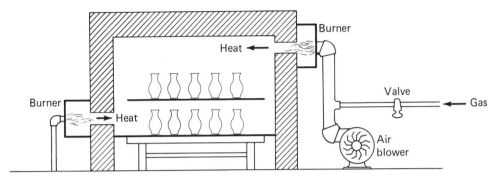

FIGURE 3-23
End view of a tunnel kiln using air and gas as fuel.

An advantage of the continuous kiln is that fuel is saved. Because of the energy crisis and cost of fuel, this is very important. The center of the kiln is kept at a constant temperature. Therefore, fuel is not used to reheat the kiln for each load. The process also produces *bisque ware* continuously. No slow-down in production takes place while waiting for the periodic cycle.

A tunnel kiln used for firing bricks may be over 250 feet long. A series of 25 or more cars, each 10 feet long, move through the kiln. The kiln shown in Figure 3-21 can fire 120,000 bricks per day. Each kiln is designed and built especially for the type and size of ware being fired.

Heating the Kiln

The type of fuel selected depends on cost and how clean it burns. Cost depends on a number of factors. First is the cost of the kiln. Tunnel kilns usually cost more than periodic kilns. The labor needed to keep the kiln in operation is the second factor. The cost of the fuel is the third factor. Some fuels are more expensive than others.

Clean-burning fuels are also a major consideration in selecting a fuel. Wood, grass, coal, coke, and oil are not clean burning. After they burn, they leave ashes or a coating of carbon. Sulfur is also produced when these fuels burn. Sulfur can damage the ceramic ware.

Propane, butane, and natural gas produce cleaner-burning flames. Natural gas is the least costly of these fuels and most commonly used. Figure 3-23 illustrates the gas method of heating a kiln.

Electricity is the cleanest method of producing heat in the kiln. Electricity causes wires in the kiln to heat, producing an extremely high temperature. A toaster is an example of how electricity is used in the kiln. Wires in a toaster heat up and produce enough heat to toast the bread. In the kiln, the heat produced is much higher than the toaster. Electric kilns are very common in the school laboratory.

Construction of Kilns

Kilns must be solidly constructed because of the extreme heat produced during the firing cycle. Insulation is used to keep heat in the kiln. Refractory bricks are used for the insulation. Bricks made of ceramic or a glass and ceramic combination withstand the intense heat in the kiln. Bricks are laid next to each other and stacked to form the inside wall of the kiln. The outside of the kiln is usually constructed of a steel case. Well-built and well-designed kilns will heat quickly with a minimum amount of fuel. During the firing cycle, the outside wall of a well-built kiln remains cool.

Temperature Measurement

Thermometers are used to measure the temperature of a room, person, or oven. For example, when cakes or cookies are baked in an oven, certain temperatures are needed for a period of time. Thermometers indicate the temperature of the oven. Ceramics also have to be *baked* (fired) for a period of time. Thermometers cannot be used with high temperatures. Most thermometers will break at 500°C (932°F). Other methods must be used to measure the temperature in the kiln. *Pyrometers* and and *pyrometric cones* are needed to measure the temperature in the kiln.

Pyrometers are instruments used to measure extremely high temperatures. Two different types of heat-resistant wires are welded together at one end. The opposite ends are placed in the instrument. Welded ends are placed in the kiln. When the wires heat up, a small electric current is produced. As the temperature increases, the amount of electricity produced is increased. Pyrometers measure the amount of electricity produced. A meter in the instrument changes the electrical measurement to a temperature measurement.

Pyrometers come in various sizes. Some are portable, while others are attached to the kiln. By checking the temperature indicated on the meter, an operator can increase or decrease the amount of heat needed in the kiln. Automatic switches will shut off the kiln when the proper temperature is reached. Figure 3–24 shows an automatic unit of a large kiln.

Pyrometric cones, also called *cones,* are small pyramid-shaped pieces of ceramic material. These cones are arranged in a series of 64. Each cone

FIGURE 3–24
Automatic instrument control panel for operating a shuttle kiln. (Bickley Furnaces, Inc.)

in the series softens and deforms at a different temperature. When the cone softens, it indicates that a certain temperature has been reached. The series begins at cone 022 and continues upward through 021, 020, to cone 01. Numbering then continues at cone 1 and continues to cone 42. Cone 022 softens at the lowest temperature. Cone 42 softens at the highest temperature. When the selected cone softens and bends, it means that the kiln can be shut off.

Different ceramic bodies have to be fired to different temperatures to reach maturity. When using cones, it is best to use three different cones. One cone represents the desired firing temperature. A second cone is one number lower. The third cone is one number higher than the desired cone. For example if cone 06 is the desired firing temperature, both an 05 and an 07 cone are also used. The lower temperature cone (07) is called the *guide* cone. A cone representing the highest temperature (05) is called the *guard* cone. All three cones are placed in the kiln. The cones are set in a *cone plaque.* A cone plaque is used to keep the cones set at the same angle. The kiln is heated slowly through the firing cycle. The guide cone will soften and bend first. This means that the temperature in the kiln is close to the desired firing temperature. The second cone will soften as the temperature increases. When the second cone bends, it means that the proper temperature has been reached. The kiln should be shut off at this time. If the third cone bends, it means the temperature is too high. Figure 3–25 shows the cones in a cone plaque. Figure 3–26 shows what happens during the firing cycle.

Cones are inexpensive and easy to use. Some kilns are equipped with an automatic cone shut-off. Small cones are used to indicate the proper temperature. A selected cone is placed on two small supports inside the kiln. Over the cone is placed a wire that is connected to the kiln shut-off switch. When the cone softens, the wire will drop down and turn off the kiln. Figure 3–27 shows how the automatic cone shut-off operates.

Cones represent different temperatures. Table 3–1 shows the temperature for each cone. The table also shows the approximate color of the kiln at different temperatures.

Types of Firing

Ceramic ware can be fired once, or more than once. After the ceramic article is shaped and dried, it is called *greenware.* Greenware is very brittle and fragile. Therefore, it must be handled carefully. Greenware may be decorated after it is cleaned. (Cleaning and decorating are discussed in the next unit.) Cleaning is the process of removing imperfections in the ceramic ware. Before greenware is fired, it must be cleaned, but it does not have to be decorated. After the greenware is fired, it is called *bisque.* Bisque is matured ceramic ware. It is a hard, strong ceramic product that can be decorated.

If bisque ware is *glazed,* it must be fired again. *Glaze* is a decorative or protective coating of glass. When glaze is applied and fired, the firing process is called the *glost* fire. In the glost fire, the glaze melts and forms the glass coating on the ceramic ware.

The type of firing determines how

UNIT 3: DRYING AND FIRING   61

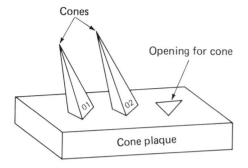

FIGURE 3-25
Cones in a cone plaque.

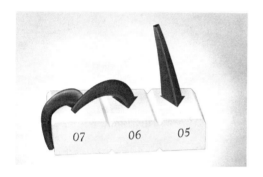

FIGURE 3-26
Guide and firing cones deformed by heat. (Edward Orton, Jr. Ceramic Foundation)

Before firing

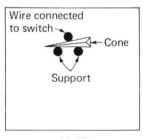

Inside kiln

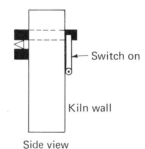

Side view

After firing

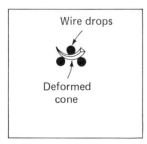

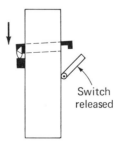

FIGURE 3-27
Automatic cone shut-off.

TABLE 3-1 CONE TEMPERATURES

Cone Number	Temperature °F	Temperature °C	Approximate Color of Fire	Type of Ware, Glazes, and Clays
15	2608	1431		
14	2491	1366		
13	2455	1346		
12	2419	1326		
11	2399	1315	White	Porcelain matures
10	2381	1305		
9	2336	1280		China bodies
8	2305	1262		Stoneware
7	2264	1240		Salt glazes
6	2232	1222		
5	2185	1196		
4	2167	1186		China glaze
3	2134	1167		Red clay melts
2	2124	1162		Semivitreous ware
1	2109	1153		
01	2079	1137		Earthenware
02	2048	1120	Yellow	White clay matures
03	2014	1101		
04	1940	1060		Low-fire earthenware
05	1915	1046		Lead glazes and low-fire
06	1830	998		fritted glazes
07	1803	983		Red clays mature
08	1751	955	Orange	
09	1693	922		
010	1660	904		
011	1641	893		
012	1623	883		Luster glazes
013	1566	852	Cherry red	
014	1540	837		
015	1479	804		Chrome red glazes
016	1458	782		Organic material burns out
017	1377	747		Overglaze colors,
018	1323	717		enamels, and gold
019	1261	682		
020	1175	635	Dull red	
021	1137	613		
022	1112	600		Dehydration

NOTE: Cones above number 15 are used for special purposes only.
Based on Orton cones—Edward Orton Jr. Ceramic Foundation.

UNIT 3: DRYING AND FIRING   63

FIGURE 3-28
Ware being stacked for bisque fire. (Wedgwood)

FIGURE 3-29
Methods used in dottling.

Tile setter   Cup supported on stilt   Plates on a plate rack

the kiln is loaded. Greenware can touch and be stacked on top of each other. After bisque is glazed, the ware must not touch each other or the kiln. If they touch when the glaze melts, the ware will stick to the kiln or other ware.

To save time in industry, some greenware is glazed and fired only once. Some decorative processes need three and more firings.

Stacking the Kiln

Kilns are loaded differently for the bisque and the glost firing. When firing bisque, the ware may be stacked on top of each other. There is no need to separate the greenware for the bisque fire. Care should be taken in stacking the ware. Excess weight should not be placed on top of other ware. Figure 3-28 shows ware being loaded into the kiln for the bisque fire.

If the greenware or bisque is glazed, they must be separated in the kiln. Separating the glazed ware is called *dottling*. Many products are available for dottling glazed ware. Figure 3-29 shows a few of the many methods used in dottling.

ACTIVITIES

1. Build Your Vocabulary:
   a. vitrifies
   b. firing cycle
   c. organic impurities
   d. oxidation
   e. saggers
   f. cones
   g. greenware
   h. bisque
2. Using forming methods described in Unit 2, produce a number of pieces of ceramic ware. Allow the ware to dry using methods described in Unit 3. Care should be taken so that the ware does not crack.
3. Practice loading the kiln for both bisque and glost firing.
4. See Unit 4 for cleaning instructions. Clean your greenware. Load the kiln and fire the ware to the proper temperature.

This unit acquaints you with the finishing and decorating of ceramic ware.

1. You will be able to describe the various methods of decorating ceramic ware.
2. You will be able to describe the various types of coatings used to decorate ceramic ware.
3. You will be able to clean and decorate ceramic ware.
4. You will be able to diagram the various decorating methods and firing techniques of ceramic ware.

In Unit 3, you learned about drying and firing ceramic ware. Recall that greenware must be cleaned before the ware is fired. However, greenware can be decorated before or after the bisque firing. This unit describes the cleaning of greenware, and the finishing and decorating methods used in ceramics.

CLEANING

Seam lines and *flaws* may be present on the ware after it is shaped. Seam lines are lines on the ware, which have an excess amount of ceramic material. They are caused by the forming process. The place where two halves of a mold join will cause a seam line. Jiggering produces a line on the product near the edge of the ware. Excess slip will run out between the two pieces when attachments are used. The slip will dry and needs to be removed. Other flaws may occur because of the forming process. Removing the seam lines and flaws is referred to as cleaning.

Seams and flaws are removed by scraping the ware with a clean-up tool. Clean-up tools are purchased in many shapes and sizes. The flaw is scraped until it is level with the surface of the object. Automatic machines are manufactured that also remove the seam

UNIT 4: DECORATING AND FINISHING

FIGURE 4-1
Turning. (Wedgwood)

lines from shaped ware. The cleaning process is usually done by hand. The entire process of removing seams and flaws is called *fettling*.

Turning is another method of cleaning greenware. Shaped ware is placed on a machine similar to a lathe. The ware is rotated past a cleaning tool. This tool scrapes off the top surface of the ware, removing any surface flaws. Figure 4-1 illustrates the turning operation.

Splash marks or small flaws caused by the forming process can also be removed by *sponging*. A damp sponge is used to smooth the flaws. After seams are scraped, they should be sponged. Only enough sponging to smooth the surface is required. If the ware is sponged too long or hard, a rough surface is produced. Sponging adds water to the ceramic ware. Before the ware is fired, it must be dried. After the object is thoroughly cleaned, it can be decorated or bisque-fired.

TYPES OF FINISHING

Ceramic ware is finished for two main reasons. *Protection* of the ware is the first reason. Fired ceramic bodies are *porous*. Porous material will absorb water. To seal the pores, the ware is coated with glaze. Glazes melt during the glost fire, thus sealing the pores. A second reason for finishing is to improve the appearance of the ware. Improving the appearance is called *decorating*.

Over the years many finishing and decorating methods have been developed. Methods of cutting or scraping the ware to create designs are often used. Minerals are added to the ceramic body to change the color. Colored slips can be brushed over the object. Colored glazes form a thin glass coating on the ware. Thin layers of metal can be fired over a glaze. These are only a few of the methods used to finish and decorate the ware.

All of the methods used can be placed into one of two groups. The first group includes methods used to change the shape or color of the ceramic body. A second group includes glazes and decorations placed over the ceramic body. Table 4-1 lists some of the more common finishing and decorating methods.

METHODS APPLIED TO THE BODY

Incising

Incising is a method used to change the appearance of the object's surface. Designs are cut into the surface of leather-hard greenware. Dry greenware is brittle and may crack during incising. Incising is a simple method of applying a design to the object's surface. A pocket knife, toothpick, fork, wire brush, and other similar objects can be used. Designs are scratched into the ware's surface. When incising is done, a small groove is cut into the surface. Grooves should be cut so that they are wider on the top. If the groove is wider at the bottom, air bubbles may form when a glaze is applied. Figure 4-2 shows a design incised on a ceramic tile.

Piercing

Piercing is a method of cutting the design completely through the leather-hard greenware. The process is similar to incising except the design is cut all the way through the wall of the ware. After the design is cut, the excess material is removed. This method is useful in decorating ware such as hanging planters, candle holders, and lamp bases. Figure 4-3 shows an object that has pierced decorations.

Inlaying

Inlaying is a method similar to incising. A design is incised into leather-hard greenware. The incising produces a series of grooves in the ware. When inlaying, the incised grooves are filled with a soft plastic clay of a different color. The grooves are overfilled, and the ware is allowed to dry. When the excess clay is leather-hard, it can be scraped off. Scraping is done to make

TABLE 4-1 FINISHING AND DECORATING METHODS

Changing the Body	Over-the-Body Methods
Incising	Stains
Piercing	Glazes
Sgraffito	Metallics
Inlaying	Decals
Slip trailing	Lusters
Underglaze	
Englobes	

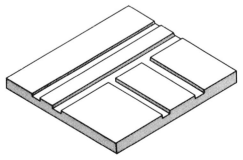

FIGURE 4-2
Incising.

FIGURE 4–3
Appearance of ware with pierced decoration. (Wedgwood)

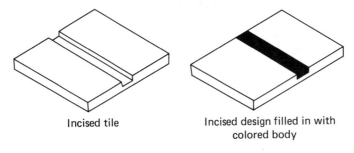

Incised tile

Incised design filled in with colored body

FIGURE 4–4
Inlaying method of decorating.

the inlayed design level with the surface. Figure 4–4 shows the inlaying method of decoration.

Relief Decorations

Relief decorations are leather-hard ceramic designs that are applied to the surface of ceramic ware. The design will be higher than the original ware. When the design is above the surface, it is called a relief. Designs used in relief work can be made in a number of ways. The design can be cut from a slab of clay. Another method is to form the design by hand molding. Finally, the design can be formed in a mold. Figure 4–5 shows a relief design being made in a mold. No matter how the design is formed, it is attached to the ware with water and slip. Designs can be made of the same or different color of ceramic body. Figure 4–6 shows a relief design being applied to the ware.

FIGURE 4-5
Relief design being made in a mold. (Wedgwood)

FIGURE 4-6
Relief design being applied to ware. (Wedgwood)

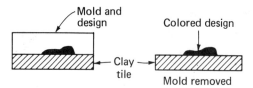

FIGURE 4-7
Sprig decoration.

Sprig Decorations

Sprig decorating is another method of applying a relief design. A design is carved into a plaster mold. Soft plastic clay is pressed into the carved design. The mold and design are then pressed against a leather-hard ceramic object. Since the ceramic body and design shrink, the design releases from the mold. Figure 4-7 shows how sprig decorations are applied.

Englobe Coating

Many ceramic objects are made from ceramic materials that have an undesirable color. Coating the ware with an *englobe* is a method used to improve the appearance. An englobe is a high-quality colored slip. Leather-hard ware is coated with an englobe and allowed to dry. After firing, the ware will have a thin layer of the desired color material on the surface. Figure 4-8 shows the englobe decorating process.

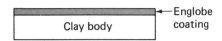

FIGURE 4-8
Englobe decoration.

Sgraffito

Sgraffito is a combination of englobe coating and incising. Shaped ceramic ware is first coated with a colored englobe. A design is incised or scratched through the englobe. The finished ceramic ware will have at least two colors. Designs will be the color of the ceramic body, showing through the colored englobe. Figure 4-9 shows the sgraffito decorating method.

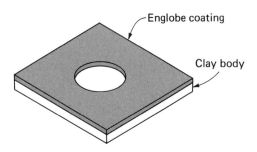

FIGURE 4-9
Sgraffito decoration.

Slip Trailing

Ceramic ware can also be decorated by *slip trailing*. Colored slip or an englobe is applied to leather-hard ware from a container. Containers such as an eye dropper or plastic squeeze bottle are filled with the englobe. Slip is then forced out of the container onto the ceramic ware. Slip trailing produces a raised design made of the englobe. Figure 4-10 shows the slip-trailing method.

Coloring Ceramic Bodies

Colorants can be added directly to the ceramic body. Various colorants are

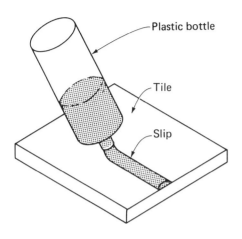

FIGURE 4–10
Slip trailing.

Underglazes

Underglazes are colored ceramic coatings that are applied to either greenware or bisque. As the name implies, they are applied under a glaze. Underglazes can be purchased in liquid, powder, and moist forms. There are two types of underglazes: opaque and transparent.

Opaque underglazes are used when a solid color is desired. Opaque means to block out. Therefore, an opaque underglaze will completely cover any color below it. If one color is applied over another color, the bottom color will not be seen. There are hundreds of opaque underglaze colors.

Transparent, or *translucent,* underglazes can be seen through. A transparent underglaze will color the ceramic ware. Any colors below the transparent color can still be seen. If a solid color is desired, transparent underglazes should not be used. A transparent underglaze can be applied over an opaque underglaze.

When the underglaze is fired, it becomes a part of the ceramic ware.

added to the slip during blunging. Different minerals are used as the colorants. After the ware is fired, the color will appear. Before firing, all the ceramic bodies look about the same. Firing causes a chemical change in the body, producing the desired color. Some of the common colorants that are used in ceramic bodies are listed in Table 4–2.

Various shades of the desired color can be produced. To obtain a deeper color, more colorant is added to the body. Usually 1 to 5 per cent, by weight, of the colorant is added.

OVER-THE-BODY METHODS

Over-the-body decorations are applied on or over the ceramic ware. They include underglazes, glazes, and overglaze techniques.

TABLE 4-2 CERAMIC BODY COLORANTS

Desired Color	Colorant
Blue	Cobalt
Brown	Manganese
Cream or ivory	Antimony
Gray	Uranium
Green	Potassium
Grayish-brown	Copper
Olive green	Black iron
Pink	Cerium
Red	Red iron
Tan	Nickel

Pores in the ware are filled with the underglaze color. However, the underglaze is porous and needs to be sealed. Glazes are used over underglazes to seal the ware, making it water resistant.

Ingredients of Glazes

A *glaze* is a thin glass coating that is fused to the ceramic ware by high heat. Glazes are made from three types of ingredients. *Silica* is the most important raw material used in a glaze. Silica's main purpose is to form the glass coating. Pure silica melts at 1700°C (3092°F). If pure silica is coated over a ceramic body and fired, the body would melt before the silica. An ingredient, called a *flux,* is added to lower the melting point of silica. The third ingredient added to a glaze is a refractory material. Refractories are used to increase the toughness and wear resistance of the glaze.

Silica is also called flint and quartz. Sand and sandstone are forms of silica. When the silica melts, it changes chemically into glass. A detailed discussion of glass is presented in Unit 5. As the silica is heated, it expands and changes shape. Because of the expansion and high melting point, it must be mixed with the other ingredients.

Fluxes are used to lower the melting point of the glaze. Oxides are the common source of fluxes. An oxide is an element that has combined with oxygen. For example, lead mixed with oxygen forms lead oxide. Oxides were formed during the rock cycle. Hot molten elements combined with oxygen to form the oxides. Common oxides are sodium, potassium, calcium, magnesium, lead, and zinc. Each oxide has a special purpose as a flux. Some oxides are used as a flux, plus adding color to the glaze.

Lead oxide is a common flux used in glazes. The use of lead in the glaze improves the color. When a glaze contains lead, it is easy to apply. The glost firing produces a bright glossy surface.

A serious health hazard is produced when lead glazes are used. Lead is a poisonous material. Individuals working in ceramic industries are exposed to the lead poison. Breathing dust containing lead can cause sickness and even death. If lead glazes are used on dinnerware or containers used for food or drink, the consumer is in danger. Laws have been passed which prevent the use of lead glazes on dinnerware. Ceramic products that will hold liquid or food use a glaze that contains no lead. Care should be observed in the school laboratory. Use only glazes that *do not* contain lead. Labels state whether or not the glaze is safe for use with food or drink.

The final ingredient used in a glaze is the refractory material. Alumina is the most common refractory material in a glaze. Alumina is a combination of aluminum and oxygen. Alumina is commonly found with other ceramic materials. Feldspar and kaolin contain alumina. Therefore, feldspar and kaolin are used as refractory materials. Refractory materials are added to increase the toughness and wear-resistance of the glaze. Hardness of the glaze is also increased by using refractories.

Types of Glazes

Glazes can be grouped by color, firing temperatures, and appearance.

Table 4-3 presents a summary of the different groupings. Glaze manufacturers make a large number of glazes in each group. Color is not included in the table, since all colors are available. Most glazes are purchased in liquid form or dry powder forms. However, glazes can be mixed from the raw materials. Many ceramic books contain formulas and mixing instructions. A variety of different glazes can be mixed in the school laboratory.

Firing Temperatures of Glazes

Glazes are developed for use on certain types of ceramic bodies. A glaze must melt at about the same temperature that the body matures. Ceramic ware expands and shrinks during the firing cycle. Glaze used on ceramic ware must also expand and shrink at the same temperature. When the glaze and ceramic body shrink and expand together, it is called *fit*. If a glaze does not fit the body, it expands and shrinks at a different temperature. Cracks in the glaze appear if the ware shrinks differently than the body. Bare spots on the ware appear when the body and glaze expand differently while being heated.

A body that matures at a low temperature needs a low-melting-point glaze. High-maturing-temperature bodies need a high-melting-point glaze. Melting temperatures of a glaze depend on the type of flux used. Fluxes for low-firing glazes are either lead or alkaline.

Lead glazes are the most common type of low-firing glaze. A lead glaze melts and matures from cone 022 to cone 05. If the firing temperature is too high, other fluxes must be added. Lead will burn out of the glaze at higher temperatures. Bright colors are produced with lead glazes.

Since lead is poisonous, its use is limited. It cannot be used on dinnerware or food containers. Soda, potash, and borax can be used as the flux in place of lead. This type of glaze is called an *alkaline* glaze. Alkaline glazes are also low-firing-temperature glazes. Bright and safe coatings are obtained with alkaline glazes.

Higher-firing-temperature glazes mature at cone 5 and higher. These glazes are used on ceramic bodies that mature at high temperatures. High-firing glazes contain both low- and high-melting-temperature fluxes. When low-melting-point fluxes oxidize, the higher ones start to melt. Calcium carbonate is a high-melting-point flux. High-firing glazes are harder and more wear-resistant than low-fire glazes.

TABLE 4-3 GLAZE GROUPING

Firing Temperatures		Surface Texture
Type	Cone Number	
Low fire	Up to 05	Gloss
Medium fire	04–5	Matte
High fire	6–9	Crystal
Porcelain	5	Crackle
		Salt
		Wood grain
Underglaze	05–6	Raku
Metallics	019	Transparent
Lusters	018	Opaque
Decals	018	Antique
		Special effects

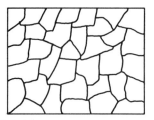

FIGURE 4–11
Crackle glaze appearance.

Frit Glazes

Frit glazes are special glazes. The glaze is melted in a container to form liquid glass. When the glaze melts, it is poured into water. As the hot liquid glaze hits the water, it cracks. Cracked glaze is then crushed into a fine powder. Glaze powder is mixed with plastic clay and used to coat ceramic ware. Frit glazes serve two purposes. First, lead glazes that are fritted are no longer poisonous. Second, all impurities are burned out of the glaze. Impurities in a glaze can cause bubbles to appear when being fired. Bubbles in a glaze can ruin the appearance of the glaze. Fritted glazes are more expensive than nonfritted glazes.

Surface Appearance of Glazes

Many glazes are developed and produced for protection and decorative reasons. Glazes can be clear or colored. *Clear* glazes are used to provide a clear glass coating over colored bodies or underglazes. *Colored* glazes are used to provide color and protection for the ware.

Glazes are also produced which have different surface appearances. *Glossy* glazes produce a shiny surface on the ware. *Matte* glazes produce a dull satin-looking finish on the ware.

A number of special glazes are used to produce different decorative effects on the ceramic ware. *Crystal* glazes have pieces of fired glaze mixed with the liquid. When the glaze is fired, the crystals create interesting designs in the surface. Another type of glaze has fine metallic powder mixed throughout the liquid. After the glaze is fired, the powder provides a sparkle in the glaze. This type of glaze resembles metallic paint on a car. Other special decorative glazes and techniques include *crackle, salt,* and *raku.*

Crackle Glaze. A *crackle glaze* has a series of thin-lined cracks in the coating. Cracks are formed to improve the appearance of the ware. When the glaze cools after firing, cracks appear darker than the glaze. Figure 4–11 shows the effect produced by a crackle glaze. Fluxes are added to the glaze, which causes uneven shrinkage. As the glaze cools, it shrinks and cracks. Because of the cracks, the glazed ware will not hold a liquid. Regular glazes will also crack if the glaze does not fit the ceramic body.

Salt Glazing. *Salt glazing* is a simple decorative process. Colorants are used in the ceramic body or in an underglaze. In salt glazing, regular table salt is used. First the ware is fired to its maturing temperature. While at this temperature, salt is thrown into the hot kiln. The salt melts and becomes a liquid. Liquid salt coats and fuses to the ceramic ware. When the ware cools, the salt forms a glasslike glaze coating.

Although the process is simple, it has a few disadvantages. The salt will also coat the inside of the kiln. Therefore, a special kiln will be needed just for salt glazing. The process also produces dangerous fumes. Adequate ventilation is a must. The third disadvantage is that only a limited number of colors can be used with salt glazing.

Raku Glazing. *Raku glazes* are low-fired glazes containing grog. Raku was developed in Japan, but is being used more and more in the United States. Decoration is the major purpose of raku. The glazed surface is rough and porous. Therefore, raku glazed ware is not waterproof. Raku is done on small simple-shaped ware. Steps in the raku process are listed below.

1. Ceramic ware is bisque-fired, then glazed.
2. The kiln is heated to the temperature needed to melt the glaze.
3. The kiln is opened.
4. The ware is placed in the hot kiln for 10 to 15 minutes.
5. When the glaze melts, the ware is removed from the kiln. The hot ware is placed in a sealed container.
6. Sawdust or wood is thrown into the container.
7. Smoke is produced and adds impurities to the glaze.
8. While still hot, the ware is removed and cooled in water. The ware will not crack when cooled.

NONFIRED STAINS

Nonfired stains are used to decorate bisque ware. Stains provide a permanent colored finish. Nonfired stains must not be fired after they are applied. Firing will cause the colors to fade and burn off. Stains are used only to color the ware. They do not protect the surface. Clear plastic or varnish is sprayed over stained ware. The coating keeps the stain from becoming dirty.

Stains are purchased in both opaque and translucent colors. Opaque stains will cover the ware with a solid color. A popular decorating method using stains is called *antiquing.* In antiquing, a translucent color is applied over the opaque color. Excess translucent colors are wiped off the ware's surface. The translucent color makes the ware appear old, or antiqued. A number of different decorative techniques are possible when using stains.

DECORATIONS APPLIED OVER GLAZES

The last group of decorations for ceramic ware are those applied over a glaze. Decorations applied over a glaze include overglazes, lusters, and metallic coatings.

Overglazes

Overglazes are colors that are applied over a glaze. Ware is first glazed and fired. Overglazes are applied over the fired glazed ware. The overglaze contains a *gum* solution. Gum is used to hold the overglaze color on the glazed ceramic ware. Low-melting-point fluxes are used in an overglaze. Therefore, the glaze must be fired first. An additional firing is needed when overglazes are used. The final firing is

FIGURE 4–12
Dipping ware. (Wedgwood)

FIGURE 4–13
Applying design with a brush. (Lenox)

done at a low temperature. Cone 016 is usually high enough for the overglaze to fuse to the glaze.

Lusters

Lusters are very thin coatings of metal that fuse into a glazed surface. Lusters are applied over a fired glaze. A fired luster has an appearance similar to a pearl. Lusters also look like a thin coating of oil on water. Lusters are produced by mixing a metallic salt with oil of lavender. Oil of lavender acts like the gum solution used with overglazes. Coated ware is fired at a low temperature to fuse the luster.

Metallics

Metallic coatings are thin layers of precious metal applied over a glaze. Common metals used are gold, silver, and platinum. The metal powder is mixed with oil of lavender and coated over the glaze. Temperatures used to fire metalic coatings are near cone 019. During the firing, the oil of lavender burns out and the metal melts over the glaze. After the ware cools, a thin, shiny metal surface is left. In addition to decorations, this method is used on many electrical and electronic products. The metal is used to conduct electric current.

METHODS OF APPLICATION

A number of methods are used to apply the protective and decorative coatings. Dipping, brushing, pouring, spraying, and stencils are the common methods.

Additional methods include stamping and decals.

Dipping of ceramic ware is a common method of applying glazes. Ware must be held carefully by skilled workers. Ware is dipped into a container of liquid glaze for a short period of time. When the ware is removed, care and skill are needed to obtain an even coating of glaze. Figure 4–12 shows a worker dipping a vase.

Liquid decorations can be applied by hand brushing. Brushing is a widely used method for applying decorations. Figure 4–13 shows a worker using a brush to apply decorations to china.

Spraying is a simple method of applying decorations. Compressors and spray guns are used to apply the liquid. Air brushes are small spray guns used to apply detailed designs. Automatic sprayers are used in production lines. Ceramic ware moves past the sprayers. An even coating of glaze can be achieved by using spraying methods. Proper ventilation and a spray booth are required when spraying.

Sponging is another decorative method used to apply a liquid colorant. A sponge is dipped into the liquid and then dabbed over the ware. Interesting decorations are produced by sponging.

Underglazes and overglazes are also made into crayons. *Crayons* are used to apply the colorant to the ware. Crayons are made from white clay and the colorant. Figure 4–14 shows the use of crayons.

Rubber stamps are also used to apply a decorative design. Designs are made of rubber and held in a holder. The stamp is pressed into the colorant. Colorant sticks to the rubber stamp. By pressing the stamp on the ware, the colorant and design are transferred to the ware. Figure 4–15 shows the rubber stamp method.

Stencils of different types are used to decorate ceramic ware. A design is drawn and cut out of wax paper or aluminum foil. The design is then placed over the ware. Colorants are brushed, sponged, or sprayed through the stencil. When the design is removed, the colorant remains on the ware. A stencil

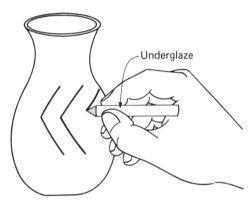

FIGURE 4–14
Use of a crayon to color ware.

FIGURE 4–15
Rubber-stamp method of applying decoration.

FIGURE 4–16
Stencil-decorating method.

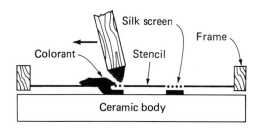

FIGURE 4–17
Silk-screen decorating.

can be reused many times. Figure 4–16 shows the stencil method of applying the design.

Another stencil method is called *silk screening.* Stencils are cut into a special paper that sticks to a piece of silk. The silk and stencil are held in a frame. Liquid colorants are forced through the screen by a squeegee. Colorants will pass through the screen not covered by the stencil. Detail stencils are produced by photographic processes. Figure 4–17 shows the silk-screening method.

Decals are another method used to apply decoration to the ware. Special paper is used to produce a decal. The paper is coated with a varnish, then the design is printed over the varnish. Instead of printing with ink, ceramic colorants are used. Decals are soaked in water to remove the paper. After the design and varnish releases from the paper, it is applied to the ware, When fired, the varnish burns off, leaving the colorant. Usually overglazes are used in decals.

Transfer paper designs are the last decorative method. The design is printed on thin paper. Ceramic colorants are used instead of ink. The paper is pressed against the ware and the design is forced onto the ware. Rubbing with your finger forces the colorant off

UNIT 4: DECORATING AND FINISHING 79

FIGURE 4–18
Design being applied by the transfer method. (Wedgwood)

FIGURE 4–19
Transfer design on a finished product. (Wedgwood)

TABLE 4-4 DECORATIVE METHODS

Sample Processes

1. GW + CG, GT, or GO → GF
2. GW → BF → CG. GT, or GO → GF
3. GW + UG → BF → CG → GF
4. GW + UG + CG → GF
5. GW → BF → UG → F → CG → GF
6. GW → BF → UG + CG → GF
7. Process 1, 2, 3, 4, 5, or 6 + M → LF
8. Process 1, 2 3, 4, 5, or 6 + L, D, or PM → LF
9. Process 1, 2, 3, 4, 5, or 6 + L or M → LF → PM → LF

Code

+ done in the same step
→ next step

GW = Cleaned Greenware
UG = Underglaze
CG = Clear Glaze
GT = Colored Transparent Glaze
GO = Colored Opaque Glaze
BF = Bisque Fire
GF = Glost Fire
F = Fire used on bisque before glost—temperature usually same as BF
LF = Low Fire 019 and 018
M = Metallic
L = Lusters
D = Decals
PM = Precious Metals (Gold, Silver, etc.)

the paper onto the ware. Figure 4–18 pictures a design being applied by the transfer method. Figure 4–19 shows the final product after firing.

In summary, Table 4–4 shows different decorative methods. The combination of different methods are endless. Only a few are shown in Table 4–4.

ACTIVITIES

1. Build Your Vocabulary:
    a. fettling
    b. incising
    c. piercing
    d. inlaying
    e. sgraffito
    f. frit
    g. crackle
    h. raku

2. Clean greenware and decorate it, using any of the first six methods described in Table 4-4.
3. Using the ware from Activity 2 (or making another object), use any of the decorating methods from seven to nine as described in Table 4-4.
4. Using ceramic tiles made from casting or slabs, try the different methods of applying decorations to the body (incising, piercing, sgraffito, and so on).
5. Make a series of tiles showing all of the decorating methods described in this unit. You may want to work with others, so that each person only has to do a few of the tiles.

This unit acquaints you with two different types of ceramic materials.

1. You will be able to describe the purpose and properties of refractories.
2. You will be able to describe the purpose and types of glass.
3. You will be able to list uses of glass and refractories.
4. You will be able to describe the forming process used to produce glass ware.

Refractories and glass are two different types of ceramic materials. Refractories are similar to clay products described in previous units. However, their purpose, properties, and use differs from clay ware. Glass is another ceramic product which differs greatly from the clay-type ceramic ware. Refractories and glass have one common characteristic. Both materials have the ability to withstand high temperatures. This unit presents a brief description of each material.

REFRACTORIES

A *refractory* is a ceramic material. Refractories differ from other ceramics because they can withstand extremely high temperatures. Refractory means that the material can resist *corrosion* at high temperatures. Refractory products come in contact with corrosive solids, liquids, and gases. Corrosive materials contain chemicals which can destroy ceramic or metal products.

Refractories are used to make ceramic products that come in contact with high temperatures. Typical products made from refractories include linings for furnaces, boilers, and kilns; nozzles for rockets; and abrasives. Without refractories, it would be impossible to make steel, aluminum, copper, cement, brick, and abrasives. Other industrial products also require extreme heat during processing. Without refractories, many industrial and consumer products would be impossible. It

Properties of Refractories

Refractories are special ceramic products that must have special properties. Melting point of the refractory is important because of their use in furnaces. Some refractories can withstand temperatures up to 2760°C (5000°F) before melting. Rather than stating melting points in degree, refractories are compared to pyrometric cones. If a refractory melts at approximately 1260°C (2300°F), it would have a *pyrometric cone equivalent* (PCE) of cone 8. Table 5-1 shows some refractories with their PCE.

Refractories have a *specific gravity* of 1.8 to 4.5 Specific gravity refers to the weight of the material compared to the weight of an equal volume of water. Water has a specific gravity of 1.0. Therefore, refractories are 1.8 to 4.5 times heavier than the same volume of water.

An important property of refractories is *thermal expansion*. Thermal expansion is the amount that a refractory expands when heated. Refractory bricks are used to line various types of furnaces. As the furnace heats, the bricks expand. The amount that the bricks expand depends upon the type of refractory and the temperature used. Knowledge of the expansion is needed when designing and constructing furnaces.

Thermal conductivity is another property related to heat. Refractories are used to keep the heat in the furnace. Therefore, refractories should have a low thermal conduction value. With low thermal conduction, the heat loss is kept at a minimum.

In addition to the previously mentioned properties, refractories must also resist *cracking* and *spalling*. Cracking and spalling occur when materials are heated and cooled. Spalling is a condition in which the surface of the refractory peels off. Usually, spalling and cracking occur because one side of the brick is hot, while the other side is cool. Therefore, one side of the brick expands more than the other side. Expansion of the refractory may cause it to crack or spall.

Classification of Refractories

Hundreds of different types of refractories are produced. Each type is mixed for special reasons related to its use. Refractories are classified accord-

TABLE 5-1 SELECTED REFRACTORIES AND PYROMETRIC CONE EQUIVALENTS

Refractory	PCE	Fahrenheit	Centigrade
Low-duty fireclay	19–29	2760°–2980°	1520°–1640°
High-duty fireclay	28–33	2940°–3170°	1620°–1740°
Super-duty fireclay	33–34	3170°–3200°	1740°–1760°
Silica	30–32	3000°–3100°	1650°–1700°
Alumina	34–38	3200°–3330°	1760°–1840°

ing to the type of raw material used. The major groups of refractories are silica, fireclay, aluminosilicate, magnesia, chromate, zirconium, carbon, and carbide.

Fireclay. *Fireclay refractories* are the most commonly produced types. These refractories are made mainly from fireclay. Fireclay refractories are not used in furnaces in which metal is melted. Molten metal and fumes cause the material to *deteriorate* (wear out) rapidly.

Most fireclay bricks are used in boilers, kilns, and fireplaces. There are many different varieties produced for various purposes. Fireclay refractories are grouped as low heat, medium heat, high heat, and super high heat. Their PCE values range from cone 15 to cone 33.

One type of fireclay brick is used mainly for insulation. This type is called *insulating firebrick* (IFB). IFB is made in a different way than regular firebrick. Sawdust or wood chips are mixed in with the clay. When the brick is fired, the wood burns out, leaving small holes throughout the brick. Air in the holes provides the insulating ability of the IFBs. When using IFB, less fuel is needed to heat the furnaces. In addition to low fuel consumption, IFBs do not spall easily.

Insulating firebricks are lighter in weight than fireclay bricks and other types of refractories. The major use is in furnaces which are often cooled and heated. IFBs are also used to line ladles that hold liquid metal.

Silica. *Silica refractories* are made mainly from silica. Their major use is in lining open-hearth furnaces that produce steel. With advances in technology, steel is being made in very-high-temperature furnaces. Therefore, use of silica brick is dropping in recent years. Silica bricks are still used to line furnaces in the glass industry and in ceramic kilns.

Silica bricks keep their strength at high temperatures. The bricks will not soften before they reach their melting point. A disadvantage in using silica bricks is that they will spall below 538°C (1000°F). Therefore, their use is limited to kilns and furnaces that operate continuously over 538°C (1000°F).

Aluminosilicate. *Aluminosilicate refractories* are a combination of silica and alumina. Aluminosilicate refractories contain at least 45 per cent alumina. As the per cent of alumina is increased, the heat resistance increases. When more than 95 per cent alumina is used, the refractory is called *corundum.*

Aluminosilicate refractories are very strong and not affected by corrosive fumes or slag. Slag is an impurity formed when metal is melted. Usually slag contains acids that destroy many refractories. Their major use is in lining furnaces that are used to make cement.

Magnesia. *Magnesia refractories* are used in open-hearth and Bessemer furnaces. Magnesia refractories contain about 85 per cent magnesium oxide. Magnesium oxide comes from sea water rather than the ground. These refractories are not affected by slag.

When lining furnaces, mortar is not used to bond the bricks. Mortar is the cement usually placed between bricks. Rather than mortar, thin steel sheets are placed between the bricks. When the furnace is heated, the steel and magnesia oxidize, forming a strong bond. The bond formed is much stronger than those made by mortar.

Chromite. *Chromite refractories* contain about 50 per cent chromic oxide. These refractories are heavier than most refractories. However, they tend to spall easily. To improve the properties, magnesia is added to chromite refractories. Higher spall resistance and lower thermal conductivity result from the magnesia-chromite combination.

Zirconium Oxide. *Zirconium refractories* are expensive because zirconium is an expensive mineral. The major use of this refractory is in crucibles. Crucibles are containers used for melting metal. Zirconium does not react with molten metals. The material also resists most fluxes and alkalis.

Carbon. *Carbon refractories* are also called graphite refractories. Carbon refractories are used to line the bottom of iron-melting furnaces. As the temperature is increased, the strength of the refractory is increased. Carbon is a useful refractory up to 3315°C (6000°F). One problem with carbon refractories is that they should only be used below the liquid metal. When used above the metal, the carbon reacts with air and will burn away. Carbon refractories are also used as electrodes in electric furnaces.

Carbide *Carbide* is a very hard refractory that is also used as an abrasive. Carbide refractories are made in electric furnaces. Silica and coke (a form of coal) are fused together at very high temperatures. Most carbides resist spalling, even at high temperatures. Carbides are used mainly as abrasives and cutting tools. They have an extremely high melting point of over 3871°C (7000°F).

Forming Methods

Refractories are formed into their desired shapes by several methods. Raw materials are mined, crushed, refined, and mixed. Forming methods include plastic molding, dry and semidry forming, hot pressing, and slip casting. One additional method of forming is also used to shape refractories. *Fusion* forming is the process in which the raw materials are melted and mixed together. While liquid, the materials are poured into steel molds and allowed to cool. This process is similar to solid slip casting, except molten materials are used.

Available Forms

Refractories can be purchased in a number of forms. Common forms available include castables, solid shapes, and plastic forms.

Castables are dry forms of the refractory. Castables contain sand, gravel, and the refractory materials. The refractory material used is usually in the form of grog. Grog is used to reduce the shrinkage after the dry form is mixed with water. Castables are more expensive and lighter weight than concrete.

Plastic forms are the same consistency as plastic clays. The plastic forms

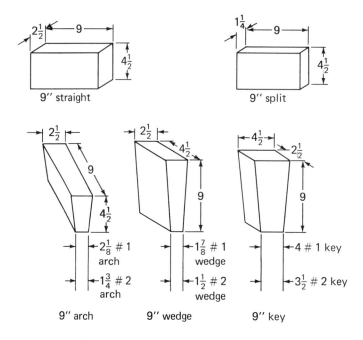

FIGURE 5-1
Standard firebrick shapes.

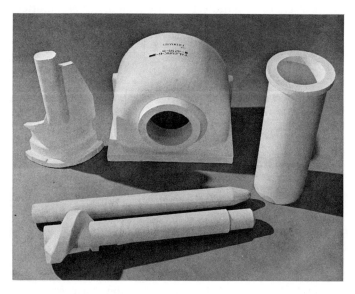

FIGURE 5-2
Special shapes of refractories. (Babcock and Wilcox)

are shaped like bricks. After being stacked along the wall of the furnace, they are pounded by hand or machine. Pounding the forms allows the wall to be shaped into one solid piece. After the wall is formed, the furnace is lit. Heat in the furnace causes the plastic form to mature.

Bricks and special shapes are matured refractories of the desired shape. Bricks come in many different forms, shapes, sizes, and types of refractories. Figure 5–1 shows some of the different-shaped bricks. Bricks are used mainly to line floors and walls of furnaces. Special shapes of refractories are also produced, such as those in Figure 5–2. Refractories are also formed into fibers and fiber insulations. Figure 5–3 shows a refractory blanket capable of insulating up to 1260°C (2300°F). Another use of refractories is in making crucibles. Figure 5–4 shows crucibles made of three different refractories.

GLASS

Glass is a widely used ceramic product. Corning Glass Works estimates that there are over 750 different types of glass and glass-ceramics available. These different glasses and glass-ceramics are used to produce over 50,000 different products.

Glass is made from silica mixed with other minerals and melted at high temperatures. Molten glass is formed into desired shapes and allowed to cool. Glass differs from other ceramic materials because it does not crystallize. *Crystallized* materials have atoms which attach themselves in an orderly manner. Water is a noncrystallized liquid. As water freezes, atoms attach themselves, forming crystals. Liquid glass is a noncrystallized liquid like water. However, when glass becomes a solid, it does not form crystals like ice. For this reason, glass is sometimes called a *hard liquid.*

Properties of Glass

Glass is a unique ceramic material. The properties of glass make the mate-

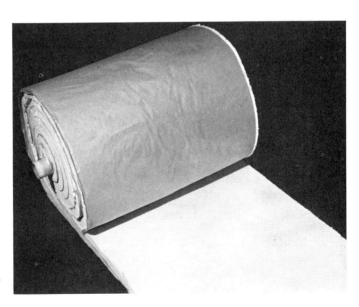

FIGURE 5–3
**Kaowool ceramic blanket.
(Babcock and Wilcox)**

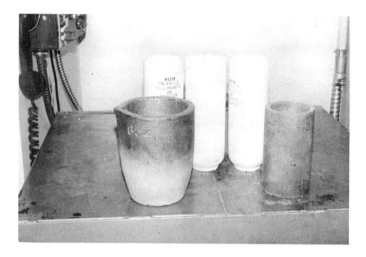

FIGURE 5–4
Crucibles. Graphite and clay on left; silicon carbide on right; aluminosilicate in the rear. (W. J. Patton, *Materials in Industry,* 1968, p. 68. Reproduced by permission, Prentice-Hall, Inc., Englewood Cliffs, N.J.)

rial very useful in our vast industrial society.

Glasses are a very brittle material. While in a brittle state, glass can break easily. There are methods used to strengthen glass. By heating glass to a predetermined temperature and cooling slowly, the glass is strengthened. Heating the glass to add strength and reduce brittleness is called *thermal strengthening.* By using chemicals, the surface of the glass is strengthened. Glass that is *chemically strengthened* has greater surface strength and compressive strength. Figure 5–5 shows a piece of chemically strengthened glass being bent.

Most glasses resist attack from most chemicals. However, hydrofluoric and phosphoric acids are two chemicals which usually cause glass to dissolve.

Glass is a hard ceramic material. When tested, a scratch test is used rather than a Rockwell or Brinell tester. Glass is rated between 5 and 7 on the Moh's hardness scale. Mild steel, aluminum, and copper are softer than glass.

The term *optical properties* refers to how clear a material is. You can see through glass unless it is colored or opaque. Glass has very good optical properties. About 90 per cent of all light can pass through glass. Objects can be seen clearly through glass.

Classification of Glass

Most of the different types of glass produced are grouped into six major types. Glasses are grouped according to the type of ingredients used and their properties. The six basic types of glass are soda-lime, lead-alkali, borosilicate, aluminosilicate, 96 per cent silica, and fused silica.

Soda-lime. *Soda-lime* glass is the most common type of glass produced. About 90 per cent of the glass produced is of this variety. Products made from soda-lime glass include bottles, windows, sheet glass, and light bulbs. This type of glass is the lowest in cost of the six types.

Ingredients in soda-lime glass consist of about 70 per cent silica, 20 per cent soda (sodium), and 10 per cent lime (limestone). When the per cent of soda is increased and silica decreased, the melting point is lower. Also the glass's resistance to chemicals is lowered. To improve the chemical resistance of the glass, a small amount of alumina is added.

Soda-lime glass cannot withstand very high temperatures. Sudden changes in temperatures may also affect the glass. This type of glass softens at about 693°C (1280°F).

Lead-Alkali. *Lead-alkali* glass is more expensive than soda-lime glass. The ingredients of this type of glass are silica 70 per cent, lead oxide 15 per cent, and soda. Lead is used instead of lime. Lead-alkali glass is very heavy, with a melting point over 621°C (1150°F).

Artware, crystal tableware, electrical components, and tubes for neon lights are made from lead glass. Lead-alkali glass is also called flint glass. It has a very high insulating ability, but cannot withstand high temperatures.

Borosilicate. *Borosilicate* glass contains about 80 per cent silica and 15 per cent boric oxide. Boric oxide is used to replace lead and soda. With the addition of boric oxide, the glass can withstand high temperatures. Borosilicate glass can withstand high temperatures. Borosilicate glass can be heated and cooled over and over without cracking.

FIGURE 5-5
Chemically strengthened glass. (Corning Glass Works)

The glass is also resistant to most chemicals.

Typical products of borosilicate glass include sealed-beam lights, telescope mirrors, laboratory table tops, furnace sight glass, and oven cookware. Pyrex, a product of the Corning Glass Works, is commonly found in the home.

Aluminosilicate. *Aluminosilicate* glass has high heat resistance, similar to borosilicate glass. However, this type of glass is very expensive. The glass softens at about 1093°C (2000°F), which makes the glass hard to form. Aluminosilicate glass has great strength and is used for various laboratory applications and high temperature thermometers.

96 Per Cent Silica. *96 per cent silica* glass contains 96 per cent silica plus boric oxide. The glass is lightweight but very strong. It resists extremely high temperatures. Because of the forming process, this glass is very expensive. 96 per cent glass is made from borosilicate glass that is heat treated and bleached in acid. Acid removes the soda from the glass, leaving silica and boric acid.

Major uses of 96 per cent silica glass include furnace sight glasses, windows on space vehicles, rocket nose cones, and glass used in the chemical industry. This type of glass can be heated till red hot, then thrown into cold water without cracking.

Fused Silica. *Fused silica glass* is made from pure silicon oxide. This is the most expensive glass produced. Fused silica glass is also called quartz glass. Fused silica glass is lightweight, but melts at very high temperatures. Maximum resistance to heat and outstanding optical properties make this glass very valuable.

Mirrors and lenses used in astronomical telescopes, windows in spacecraft, and laser beam reflectors are only a few products made of fused silica glass.

Special Glasses

In addition to the six major types of glass, there are a number of special glasses. A few of the many special glasses are colored, optical, photochromic, fibrous, cellular, and glass-ceramics.

Colored glass can be produced by adding various chemicals to the batch. Glass has a natural green-blue tint. Chemicals are used to remove this color from the glass. Some of the chemicals used to change the color of glass are shown in the following table.

Chemical	Effect
Iron	Green
Manganese	Violet
Cobalt	Blue
Copper chloride	Red
Neodymium	Yellow
Tourmaline	Polarized glass

Optical glass can be made from any of the six types of glass. While the glass is being melted, bubbles which appear are removed. Removing the bubbles makes the glass clear and satisfactory for eyeglasses.

Photochromic glass will turn darker when exposed to ultraviolet

FIGURE 5-6
Photochromic sunglasses. (Corning Glass Works)

light. When the light is removed, the glass will lighten to the original tint. Eyeglasses and sunglasses made from photochromic glass can now be purchased. Figure 5-6 shows the darkening and fading action of photochromic eyeglasses.

Fibrous (fiber glass) glass is made into thin fibers. Fibrous glass is made by melting the glass and then forcing it through a small hole, or die. Pressure is used to force the glass through the die. This process is similar to using a piston-type extruder. Extrudants are very thin fibers of air and glass. Fiber glass is used as an insulation in the home. When the glass fiber is mixed with plastic, it forms a tough material used in furniture, skills, fishing rods, and car and boat bodies.

Glass produced by mixing powered glass with a foaming material is called *cellular glass*. This mixture is heated to melt the glass. When the liquid glass is cast into shape, the products have many small cells throughout the material. Cellular glass is lightweight

FIGURE 5-7
Machining cellular glass. (Corning Glass Works)

and can be machined with ordinary tools. Figure 5-7 shows cellular glass being machined on a vertical milling machine. Cellular ceramics are being

FIGURE 5-8
Cellular glass shapes. (Corning Glass Works)

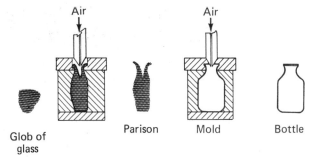

FIGURE 5-9
Blow-molding glass.

used in catalytic converters to reduce air pollution from cars. Figure 5-8 shows some of the many shapes available in cellular glass.

When glass is melted with chemicals added to form crystals, a *glass-ceramic* is produced. Recall that glass does not have crystals. When crystals are formed, a new type of product is developed. Glass-ceramics are harder than glass, stronger, and can withstand extreme temperatures. Common products made from glass-ceramics include cookware, rocket nose cones, flattop cooking stoves, tape recorder heads, and special electronic components.

Forming Methods

Raw materials are refined in a method similar to that used for other ceramic products. Glass materials are mixed and melted in furnaces made of refractory materials. Furnaces come in a number of sizes. Some furnaces are capable of melting over 1,000 tons of glass at one time.

Glass is formed by pressing, blowing, and drawing. *Pressing* is similar to the pressing methods described in Unit 2. Molds used in pressing are made of special metals rather than plastic. A glob of hot glass is placed in the mold and pressed into shape. Most flat- and shallow-shaped ware is manufactured by pressing.

Blowing methods are used to form bottles and deep-shaped objects. Blowing consists of two steps. First, the glass glob is formed into a temporary shape called a *parison*. The second step is to hold the parison in a mold and blow air into the parison. Air pressure causes the parison to take the shape of the mold. Figure 5-9 shows how blow molding is done by machine. Blow molding can also be done by skilled workers. Figure 5-10 and Figure 5-11 show workers making crystal glassware by hand methods.

Drawing is a process in which *molten* (plastic state) glass is pulled through a die opening. This process is used to form special shapes. Flat sheet glass can be produced by drawing. Glass can also be forced between a series of *rollers*. The rollers force the molten glass into the desired shape. Figure 5-12 shows the drawing and the rolling methods used to form glass.

Finishing Glass

Glass that is formed may have rough edges and seams. In order to remove the seams and rough edges, flames are used to smooth the surface. Glass objects can also be ground or cut on grinding wheels. Figure 5-13 shows a burn-off machine, while Figure 5-14 shows a skilled operator cutting a design in fine crystal glass.

After the glass is polished, it is passed through an oven called a *lehr*. Lehrs are used to heat and cool the glass slowly. This process removes any stresses that may have set up in the glassware.

Special decorations are silk screened on glassware. Automatic machines are used to print the designs on glassware. Figure 5-15 shows an automatic silk-screen machine printing a three-color design on glass tumblers.

FIGURE 5-10
A skilled worker blowing a bubble. This is the first step in making crystal. (Lenox)

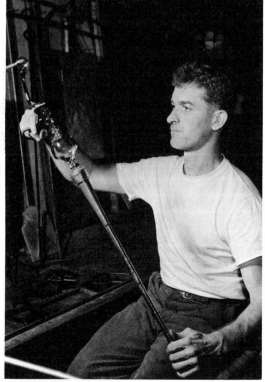

FIGURE 5-11
Forming a base on crystalware. (Lenox)

UNIT 5: REFRACTORIES AND GLASS   95

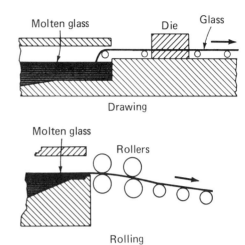

FIGURE 5-12
Drawing and rolling glass.

FIGURE 5-13
A tumbler burn-off machine.
(Eldred International
Corporation)

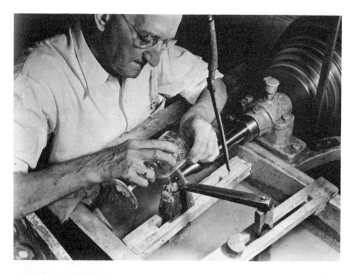

FIGURE 5-14
Cutting designs in crystalware. (Lenox)

FIGURE 5-15
Automatic multicolor decorating machine. (Eldred International Corporation)

ACTIVITIES

1. Build Your Vocabulary:
    a. refractory
    b. carbide
    c. fusion
    d. crystallized
    e. borosilicate
    f. photochromic
    g. blowing
    h. lehr
2. Develop a chart showing the different types of glass and their uses.
3. Secure samples of different refractories used in the school laboratory and develop a display.

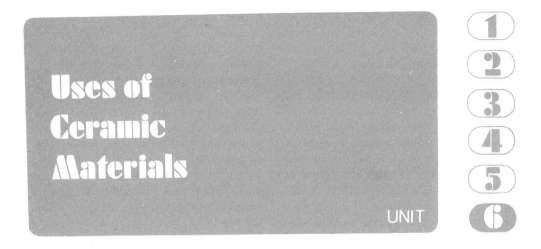

Uses of Ceramic Materials

UNIT 6

This unit acquaints you with ceramic products used in building construction, in homes, and for industrial and other miscellaneous applications.

1. You will be able to describe the difference between cement, concrete, and plaster.
2. You will be able to list and describe some of the ceramic products used in the home.
3. You will be able to list some of the ceramic products used in electrical and electronic components.
4. You will be able to list some of the ceramic products used in industrial applications.
5. You will be able to list some of the ceramic products used in miscellaneous fields.

Ceramic products are used in virtually every building, industry, and scientific field. There are hundreds of thousands of ceramic products manufactured in the United States. Ceramics are the backbone of our industrial and technological society. This unit briefly lists and describes some of the many ceramic products in use.

CERAMICS IN BUILDING CONSTRUCTION

Ceramics are used extensively in building construction. Cement, concrete, plaster, brick, tile, and sanitary ware are all ceramic products.

Cement

Cement and concrete are often confused when referring to uses of each material. Cement is an adhesive. Its major purpose is to coat and form a bond for other materials. Two major types of cement are used in building construction. The two types of cement are lime and gypsum. Lime, also called *calcium cement,* is used to form concrete. Gypsum cements are used in plaster of

Paris, plaster wall finishes, and gypsum wallboard.

Lime comes from limestone and is called calcium carbonate. Limestone contains impurities which must be removed before it can be used in cement. Removing the impurities in lime is called *calcining*. Crushed limestone is heated to approximately 927°C (1700° F). Chemical reactions occur during the heating to produce pure lime. Lime is mixed with other materials to produce cement, mortar, plasters, and soil conditioners for agricultural use. Mortars are pastelike materials used to adhere building blocks and bricks.

Portland cement, the most common type, is made from clay and lime. Each material is crushed, measured, and mixed into the proper proportions. Mixing of raw materials can be done while dry or wet (mixed with water). The mix is then loaded into a rotary kiln and burned.

Kilns used to burn the mixture are long cylinders that rotate. The kilns are very large industrial-type furnaces. Kilns are from 3 to 18 feet in diameter and from 200 to 300 feet long. This cylinder is tilted at a slight angle so that material inside will tumble and move to the lower end.

Raw materials are fed into the high end of the kiln. Heating and rotating begin, causing the material to heat and move to the low end. During the movement, a number of reactions occur, similar to the firing cycle used in clay materials. Water is removed during the low-heat periods in the kiln. Chemical changes take place during the high-heat periods. Lime in the kiln changes chemically into *di-calcium silicate* and *tri-calcium silicate*. When the material reaches the low end of the kiln, the chemical changes are complete. During the heating process, the materials fuse together, forming larger lumps. The lumps of material are called *clinkers*. Clinkers are immediately loaded into another kiln where cooling takes place. Clinkers are then crushed into a fine powder, approximately a 300-mesh size. This powder is then mixed with small amounts of gypsum and other materials.

Portland cement contains lime, silica, alumina, magnesia, soda, iron oxide, and sulphur. Seventy-five per cent or more of the cement contains calcium in the form of di- and tri-calcium silicates. Properties of the cement are determined by the percentages of other added materials. Alumina is kept to a minimum to increase strength. An excess of pure lime will cause the cement to set slowly. Lime will also cause the cement to be weak.

Cement is mixed with water to form a plastic body. Cement absorbs water and begins to set up or cure. By absorbing water, cement becomes a hard mass. When a material sets by absorbing water, it is called *hydration*.

Concrete

Concrete is a mixture of cement, water, and an aggregate. *Aggregates* are crushed rock, gravel, or sand. About 75 per cent of concrete consists of an aggregate. The remaining 25 per cent of concrete consists of cement and water in a paste form. Recall that cement is an adhesive. Therefore, the cement paste should surround the aggregate to insure a good bond. Various sizes of aggregate are used to fill in spaces between larger pieces of aggregate.

Concrete is mixed with various

amounts of water, aggregate, and cement. The amount of water has an effect upon the properties of the concrete. Less water in the mixture will produce a stronger concrete. However, concrete sets by hydration and needs water to cure properly. Concrete used to make a sidewalk, floor, or roadway is poured into position and compacted. Concrete must cure slowly; otherwise, its strength is reduced. Sun and wind may cause the surface of the concrete to dry too fast. To prevent weakness, water should be sprayed on the concrete while it cures. A plastic cover over the concrete aids in slowing the drying process.

Use of Cement and Concrete

Cement is mixed with water and sand to form a paste called *mortar*. Mortar is the adhesive used to hold blocks and bricks in place. The white or gray material between bricks on a house is called mortar.

Concrete has a number of uses in building and road construction. A list of some of the many uses follows.

- Sidewalks
- Porches
- Roads
- Parking lots
- Airport runways
- Concrete floors
- Walls of skyscrapers
- Bridges
- Walls of dams
- Ornamental birdbaths and fountains

Concrete is high in compressive strength, but low in tensile strength. To increase the tensile strength, steel rods are embedded in the concrete. Rods are placed in the concrete used for roadways and large buildings. The product is called *reinforced concrete.*

Concrete shapes can be cast at the manufacturer. Rods are added, and curing is done in a manner to improve strength. Cast shapes are then delivered to building sites and set in place. Shaped concrete forms are called *precast* forms.

Terrazzo is a mixture of cement and marble chips. Marble acts as the aggregate. Terrazzo is most often used to produce colorful floors in homes and office buildings.

Plaster

Plaster is a ceramic material made from gypsum. Gypsum is a white rock containing water. In making plaster, the water must be removed. After mining and crushing, the gypsum is heated in a kiln and calcinated, which is the process of heating the gypsum to remove the water. After the water has been removed, the material is called *plaster of Paris.* Plaster of Paris is used in wallboards, sheathing, and molds. Molds are used for slip casting of clay.

Other types of plaster are made by heating the gypsum to a higher temperature and adding additional materials. Plasters set after adding water and mixing. Plasters absorb the water and eventually dry into a hard material.

Other Ceramics Used in Construction

Common bricks are the basic structural bricks used in construction. Footings and back-up walls in buildings are the major use. Common bricks are made by either the stiff-mud or dry-press process. These bricks are not produced for surface appearance. For appearance, face bricks are used.

Face bricks are denser and more colorful bricks. Their major use is to form the outside surface of walls. The brick is more resistant to weather conditions. Face bricks are made with different surface textures. Next time you are near a brick building, take a look at the bricks. Notice the color and surface texture of the bricks. Figure 6-1 shows some of the surface textures used on face brick.

Glazed structural tile is made from fireclay and silica. Tiles are glazed to produce a wear-resistant and water-resistant surface. Tiles are produced by pressing or extrusion. While the tiles are wet, a glaze is applied to the surface. Glazes must fit the body of the brick; otherwise, cracks may result.

Glazed structural tiles are commonly used in schools, hospitals, public buildings, food processing, sewage, and water treatment plants. The tiles provide a glazed surface which can be easily cleaned. Figure 6-2 shows the use of glazed structural tiles in a public building. There are a number of colors, shapes, and sizes available. Figure 6-3 shows some of the shapes that can be purchased.

Drainage tile is ceramic pipe used to drain water away from buildings. The tiles are porous, which allows them to absorb water. Tiles are laid be-

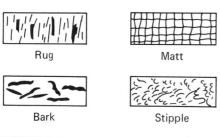

FIGURE 6-1
Surface texture of bricks. (Reston Publishing Company, Inc., a Prentice-Hall Company, Reston, Virginia 22090)

FIGURE 6-2
Use of glazed brick. (Arketex Ceramic Corporation)

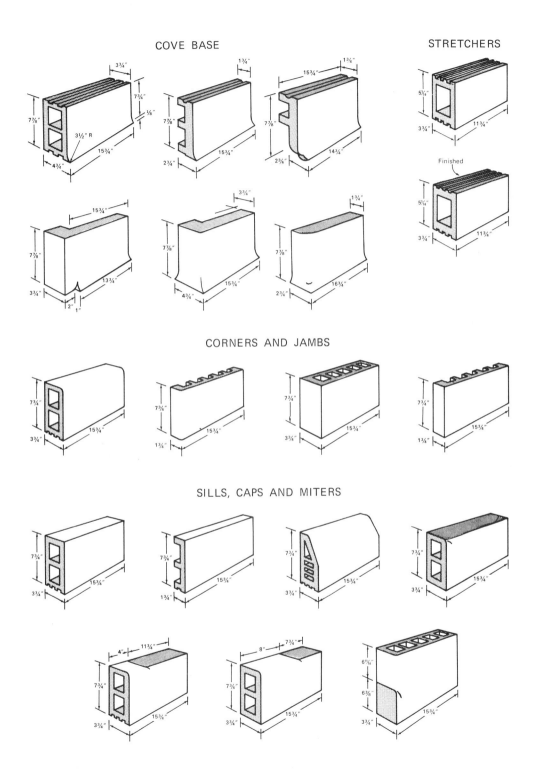

FIGURE 6–3
Shapes of structural facing tile. (Arketex Ceramic Corporation)

FIGURE 6–4
Mozaic design on a building.
(George Houlton of Photo
Researchers, Inc., New
York City)

low the ground and fitted together, forming a continuous pipe.

Clay products are also shaped and used as *roofing* material on homes and commercial buildings. The same material is used to make flower pots.

Ceramic wall tiles are thin ceramic tiles made from a porous clay body. Wall tiles are glazed to make them water resistant. Shapes and sizes are limited. Usually they are square or rectangular. However, a large number of colors are available. Ceramic wall tiles are placed on a wall with an adhesive. *Grout,* a plasticlike material, is used to fill in the spaces between the tiles. Wall tiles are used in washrooms and behind stoves and sinks in kitchens.

Mosaic tiles are small, various-shaped colored tiles. Mosaics are placed in position to make interesting designs of various colors. Mosaic tiles are used as decorations on buildings, table tops, ash trays, and so on. Figure 6–4 shows a mosaic design on the outside of a building.

Ceramic tiles are also used as floor

coverings. *Floor tiles* are made of a dense ceramic body. Various colors are available and are usually added to the ceramic body. Glazing is not used because when wet, glazed tiles are slippery. To prevent slipping, abrasives are added to the ceramic body. The major uses of floor tiles are in public buildings, terraces, porches, and kitchen floors.

Sanitary ware, such as wash basins, tubs, showers, and toilet bowls, are made of ceramic materials. A porous body is glazed with a white or colored opaque glaze. A denser body, which is less water absorbent, is used to produce ware for commercial use. The denser body is called *vitreous chinaware.*

Other ceramic materials used in construction include windows made of glass, insulation made from fiber glass, and refractories to line furnaces.

CERAMICS IN THE HOME

Many consumer goods in the home are produced from ceramic materials. Electrical and electronic products contain ceramics. These products are presented later in this unit. Other products commonly found in the home include dinnerware, figurines, table tops, cookware, crystal glassware, and fiber glass.

Stoneware is a table dinnerware having a glazed surface. Usually the base of the stoneware is unglazed. Stoneware is a thick, heavy dinnerware having various colors and designs. *Earthenware* is used for artware and dinnerware. China and ball clays are used to form earthenware. Earthenware for general purpose use is available in many shapes and colors.

China is a common material used for dinnerware. Different types of china include semivitreous, hotel, and bone. *Semivitreous china* is the most common type of china used in dinnerware. It is relatively inexpensive tableware having good strength. *Hotel china* is an opaque white vitrified ware of high strength. This china comes in various weights based on the thickness of the material. Strength is important because of its use in restaurants. *Bone china* is the pure white type of chinaware. It is highly resistant to chipping and almost translucent. Bone china is used mainly for fine tableware, ornaments, and figurines. Figures 6–5 to 6–7 picture various designs available in fine bone china. Other dinnerware is produced from porcelain. *Porcelain* is a high quality, dense, water-resistant body. Dinnerware made of a glass ceramic is made by Corning Glass Works. The material has the trade name Correll.

Vases, ash trays, planters, figurines, souvenirs, mugs, steins, bottles, and lamp bases are a few of the ceramic products used in the home. Figure 6–8 shows a vase with relief designs. Christmas tree ornaments are made of glass. There are many shapes and colors available. Figure 6–9 shows glass Christmas tree ornaments.

Glass and glass-ceramics are used for flat cooking surfaces. Warming trays are made of glass with heating elements embedded in the surface. Figure 6–10 shows a table-top cooking surface. Note the heat-resistant glass window on the oven door. Ceramic wall tile surrounds the oven for decoration. Heat-resistant glass and Pyrex

104 UNIT 6: USES OF CERAMIC MATERIALS

FIGURE 6–5
Bone china—Kutani Crane design. (Wedgwood)

FIGURE 6–6
Bone china—Caernarvon and Harlech design. (Wedgwood)

FIGURE 6–7
Bone china—Corn Poppy design. (Wedgwood)

FIGURE 6–8
Vase with relief design. (Wedgwood)

measuring cups are in Figure 6–11. Pyrex bowls and trays are also manufactured for cooking foods in the oven.

Another common ceramic material found in the home is *fiber glass.* Fiber glass is used in materials for drapes, clothes, rugs, and furniture coverings. When fiber glass is mixed with plastic, it is used to make awnings, skiis, boats, storage sheds, car bodies, and room dividers.

Graphite and clay are crushed and mixed together. The mixture is shaped into long thin rods and fired. This rod is then cut to desired lengths and set in wood. The wood is varnished and an eraser is attached. This ceramic product can be sharpened to a point. The material comes in various hardnesses. Have you guessed what the product is? Yes, it is a lead pencil. Lead pencils do not contain lead. The name comes from products formerly made of lead. Lead has been replaced by graphite and clay.

FIGURE 6–9
Glass Christmas ornaments. (Corning Glass Works)

FIGURE 6–10
Smoothtop cooking surface. (Corning Glass Works)

106 UNIT 6: USES OF CERAMIC MATERIALS

FIGURE 6-11
Pyrex measuring cups. (Corning Glass Works)

TABLE 6-1 ELECTRICAL USES OF CERAMICS

Brushes in motors	Antennae
Outlet boxes	Recording heads (tape
Switch boxes	and video tape)
Plugs	Generators
Sockets	Alternators
Fuse holders	Microphones
Fuses	Sonar systems
Insulators	Radar
Transformers	Ultrasonics
Spark plugs	Television
Resistors	Radio
Vacuum tubes	Toasters
Supports for	Telephones
heating elements	Microwave
Capacitors	Search lights
Condensers	Batteries
Light bulbs	Arc lights

ELECTRICAL AND ELECTRONIC USES

Ceramics play an important role in electrical and electronic components. Ceramics act as an excellent insulator. Most ceramic parts are made from porcelain, glass, glass-ceramics, and other clay bodies. There are thousands of ceramic products manufactured for electrical use. Table 6-1 lists some of them.

Ceramics are used in microcircuits. They are compact in size, low cost, and have very high performance abilities. Small units are used in video equipment and computers. Figure 6-12 shows a small video delay module.

Instruments are used to measure the amount of impurities in streams, rivers, and lakes. Data from the instruments is useful to agencies that enforce antipollution regulations. This instrument is pictured in Figure 6–13. Another instrument made of ceramic materials measures the pH value of liquids. Figure 6–14 shows a portable pH meter.

Fiber optics devices are manufactured from flexible glass rods. By using fiber optics, the level of water, oil, and fuel can be measured. The system has great potential for use in automobiles as a warning system for low liquid levels. Figure 6–15 shows how the system works. The sensor and warning light are shown in Figure 6–16.

Ferrites are ceramic nonmetal magnets. Ferrites are used in miniature transformers, antennae, and other electronic parts. Table 6–2 lists a few of the many uses of ferrites. Figure 6–17 shows parts made of ferrite ceramics.

Ceramics are also used in permanent magnets. Permanent ceramic magnets are used in loudspeakers, motors, generators, alternators, toys, ignition systems, and door latches. Figure 6–18 shows permanent ceramic magnets.

ABRASIVES AND CUTTING TOOLS

Many ceramic raw materials are used as *abrasives.* Abrasives are either natural or synthetic. Natural abrasives include emery, feldspar, quartz, and diamonds. These raw materials are mined and crushed, then sized. Synthetic abrasives include silicon carbide,

FIGURE 6–12
Ceramic (glass) video module. (Corning Glass Works)

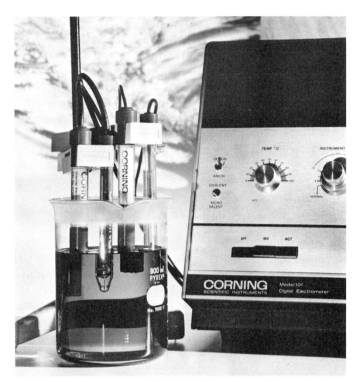

FIGURE 6-13
Pollution-monitoring instrument. (Corning Glass Works)

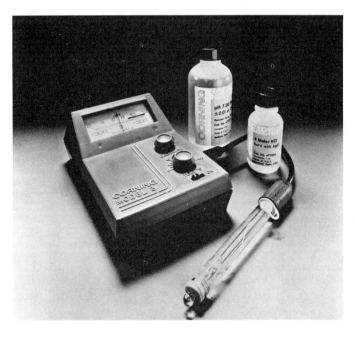

FIGURE 6-14
Portable pH meter. (Corning Glass Works)

UNIT 6: USES OF CERAMIC MATERIALS 109

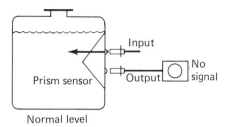

Normal level

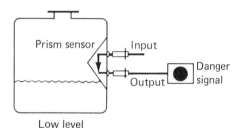

Low level

FIGURE 6–15
Fiber optics warning system. (Corning Glass Works)

TABLE 6-2 USE OF FERRITES (NON-PERMANENT MAGNETS)

Transformers in TVs	Aircraft and marine radios
Antennae	Video tape recorders
Electronic auto ignitions	Computers
Fluorescent lights	Push-button telephones
Mobile radios	Light dimmers
High frequency welders	Amplifiers used in lunar module (LEM)

alumina, artificial diamonds, and boron nitride.

The abrasives are crushed and bonded to paper or cloth. Grinding wheels are made by dry pressing or casting. Binders are used to hold the abrasive materials together. Common binders include glass-ceramics, porcelain, and plastic resins. Abrasives and binders are mixed and shaped, then fired to fuse the materials. Hundreds of shapes and sizes of grinding wheels are available.

Ceramic cutting tools are also made because of their strength and heat resistance. Ceramic and carbides are the commonly used materials. The tools are used for drilling, lathe cutting, milling, and saws. To reduce the cost of the tools, only small pieces of the ceramic material are used. Small pieces of material are soldered or brazed on the metal tools.

FIGURE 6–16
Fiber optics: sensor on the left, warning light on the right. (Corning Glass Works)

FIGURE 6-17
Ferrite ceramic parts. (Indiana General, Valparaiso, Indiana)

FIGURE 6-18
Ceramic magnets. (Indiana General, Valparaiso, Indiana)

OTHER USES OF CERAMICS

Ceramics are finding increased use in the medical and dental fields. Broken bones that cannot be repaired can now be replaced by bones made of ceramic materials. Dentures and caps used by dentists are made of porcelain. Porcelain can be colored to resemble natural teeth. In addition, it resists food acids and is wear resistant.

Lights used in operating rooms use glass reflectors that are coated with a film to separate heat from light. Heat is reflected away from the operation, while shadow-free light illuminates the area. Figure 6–19 shows the use of reflectors in an operating room.

Glass is used in the chemical laboratory because of its acid-resistant properties. Figure 6–20 shows a solvent recovery plant that handles chloroform mixtures. Figure 6–21 shows some of the many different shapes of glass tubes and instruments used in chemical laboratories.

Aerospace missions have also increased the use of ceramics. Electronic communications, computers, switches, rocket nozzles, and firing chambers are made of ceramic materials. Figure 6–22 shows an exploded view of a thermoelectric generator. The generator has two zirconium oxide ceramic rings supporting a nuclear fuel capsule. This unit provided power for Pioneer 10, the spacecraft used for the exploration of the planet Jupiter.

Metal can be coated with porcelain enamel glazes. The glaze is fritted and applied to clean metal. Fusing the glaze to the metal needs temperatures up to

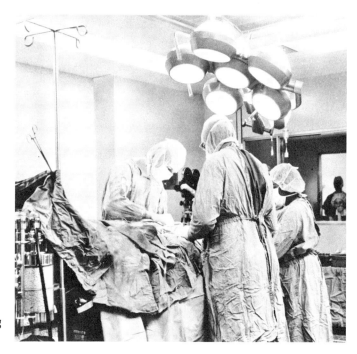

FIGURE 6–19
Reflector glass. (Corning Glass Works)

FIGURE 6–20
Solvent recovery plant—borosilicate glass. (Corning Glass Works)

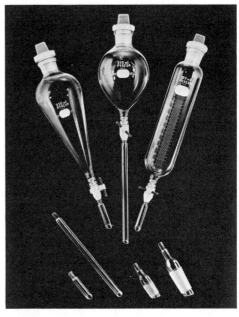

FIGURE 6–21
Chemical glassware. (Corning Glass Works)

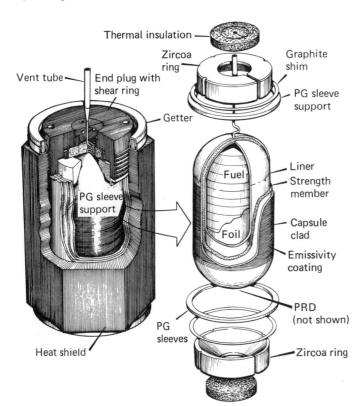

FIGURE 6–22
Thermoelectric generator for Pioneer 10 spacecraft. (Corning Glass Works)

FIGURE 6-23
Porcelain-enamelled stoves. (W. J. Patton, *Materials in Industry*, 1968, p. 67. Reproduced by permission, Prentice-Hall, Inc., Englewood Cliffs, N.J.)

816°C (1500°F). Porcelain enamel is applied to many household appliances. Examples are stoves, refrigerators, washers, dryers, cabinets, can openers, and many others. Coatings protect the metal from oxidation and add beauty to the product. Figure 6-23 shows porcelain-enamelled stoves.

ACTIVITIES

1. Build Your Vocabulary:
    a. cement
    b. lime
    c. calcining
    d. Portland cement
    e. hydration
    f. aggregate
    g. ferrites
    h. abrasives
2. Develop a list of the ceramic products found in the home.
3. Obtain cement and various aggregates, mix, and cast into small blocks. Allow the concrete shapes to harden and cure. Test each block for hardness, compression strength, and tensile strength. Record your results and determine which mixture has the best properties.

Glossary

abrading A method of comminution whereby the particle size is reduced by friction or scraping action.
aggregate Crushed rock, gravel, or sand used as a filler in concrete.
alkaline A mineral obtained from the ashes of plants. Used in place of lead in a glaze, making the glaze safe for use on dinnerware.
alumina Combination of aluminum and oxygen, usually found in nature as aluminum oxide.
alumina clay A clay containing aluminum oxide. Alumina clay has high strength and good heat resistance.
aluminosilicate A clay body containing alumina and silica, with a high resistance to heat.
antiquing Decorative process used to produce an old-looking finish.
ball clay Clay used to produce products having great strength.
batch dryers Ceramic dryer used to dry one group of products at a time.
binder Chemicals added to semidry and dry bodies, used to hold the various ceramic particles together while forming.
bisque Fired ceramic ware.
blunging Process of mixing raw materials and water to form a ceramic body.
bone ash Crushed animal bones used as a flux.
borosilicate Type of glass containing silica and boric oxide.
calcining Method of removing impurities from lime by heating.
carbide Silica and coke fused together, forming a very hard refractory material.
casting Forming method in which liquid clay (slip) is poured into a plaster mold. Water is absorbed by the plaster leaving the desired shape.
cellular Material having a series of air bubbles of air cells.
cement Ceramic material used as an adhesive to bond other ceramic (natural) materials.
ceramic body Mixture of ceramic raw materials in various proportions.
ceramics Articles made from naturally occurring materials. Raw materials are mined, crushed, refined, mixed, shaped, dried, and fired to form a ceramic product.
chemical strengthening Method used to strengthen glass.
coil building Construction of clay objects using thin coils of rolled clay.
comminution Crushing and grinding of large particles into smaller ones.
concrete Mixture of cement, water, and aggregate.
conduction Method of heating by direct contact.
cones See Pyrometric cones.
consistency The appearance, feel, and workability of a material.
convection Heating by hot air.
Cornish stone Stone found in England, used as a flux in ceramic ware.
crackle Glaze having a series of thin line cracks in the surface.
crystal Glaze having the appearance of crystals in its surface.
deflocculants Chemicals used to make slip liquid without great amounts of water.

draft Method used to circulate heated air in a kiln.
englobe Colored slip applied to the surface of ceramic ware.
erosion Continuous process of wearing away the earth's surface.
extrusion Machine-forming process used to produce a continuous product having the same cross-sectional shape.
feldspar Common type of flux, found mixed with quartz or granite.
fettling Process of removing seam lines and flaws from greenware.
fibrous Fine thin fibers of glass used as insulation or reinforcement in plaster.
fire clay Dark-colored clay used in heat-resistant ceramic products.
fit Term used to indicate how a glaze and ceramic body expand and shrink at various rates.
flow The ease with which a liquid can be poured.
flux Material added to ceramic bodies to bond the various particles together. Also used to lower the melting point of glass and glazes.
frit Glaze made from prefired glaze. Primary purpose is to make lead glazes nonpoisonous.
froth flotation Refinement process used to remove impurities from raw materials.
geology Science concerned with the study of the earth, its formation and change.
glass Mixture of silica and other materials producing a noncrystallized solid.
glaze Thin glass-like coating fired to ceramic ware, providing a decorative and protective coating.
glost Firing of a glaze; usually a second or third firing.
greenware Unfired ceramic-shaped objects.
grog Crushed, fired ceramic ware used as a filler in ceramic bodies to reduce shrinkage.
gypsum Refined rock (gypsum) used to make cement.
humidity Measure of water content in the air.
igneous Rock formed from lava and magma (mantle part of earth's composition).
incising Method of cutting a design into the surface of leather-hard greenware.
inlaying Design incised into leather-hard greenware. The design is filled with a different-colored ceramic body.
jiggering/jolleying Automated processes used to shape the inside/outside of ceramic products.
kaolin Purest type of clay, sometimes called China clay.
kiln High-temperature oven used to fire ceramic ware.
leather-hard Clay body, after some water is removed, allowing the clay to be handled without deforming.
lime Refined limestone used in mortars and plasters.
magnetic separation Use of magnets to remove metallic impurities from ceramic raw materials.
matte Dull satin-looking finish.
mesh size Number of openings per inch in a screen.
metamorphic Rock formed by long periods of heat and pressure below the earth's surface.
mortar Mixture of cement, sand, and water in a paste form, to be used as an adhesive for building bricks and blocks.
muffle Separate sealed firing chamber in a kiln.
nepheline syenite Flux used in place of feldspar. A low-melting-point flux.
opaque Solid-color coating.
overburden Top layer of earth's surface, which is removed in open-pit mining.
overglaze Decorative colors applied and fired over a glazed piece of ceramic ware.
oxidation Chemical reaction occurring during firing cycle, as impurities burn out.

oxide Element or mineral combined with oxygen.
parison Temporary shape of molten glass used in the blowing process.
piercing Design cut completely through leather-hard greenware.
porcelain Hard, dense ceramic product.
porosity Measure of the closeness of ceramic particles.
Portland cement Common type of cement made from clay and lime.
potter A craft person having great skill in making ceramic objects on a rotating wheel.
pug mill Machine used to mix and de-air clay.
pyrometer Instrument used to measure extremely high temperatures.
pyrometric cones Small pyramid-shaped pieces of ceramic material that soften at different temperature to indicate the temperature inside a kiln.
radiation Form of heating by reflected heat.
raku Low-fired glaze containing grog, and placed in a reduction atmosphere to produce a decorative effect.
refinement Process of removing impurities from ceramic materials.
refractory Ceramic material capable of withstanding high temperatures.
reinforced concrete Concrete having steel rods embedded in it to increase strength.
relief Raised design applied to ceramic ware.
residual Rock and clay found at the place where they were formed.
saggers Fired ceramic containers loaded with ware for firing.
salt glazing Decorative process where salt is thrown into a hot kiln, forming a protective coating.
sanitary ware Ceramic products used as wash basins, tubs, showers, and water closets.
sedimentary Rock and clay deposits found at a point different from where they were formed.
sgraffito Incising a design into an englobe coating on ceramic ware.
shrinkage Reduction in size of a ceramic object caused by water loss.
silica Main ingredient in clay and glass products.
slake Period of time when dry ingredients are allowed to absorb water before mixing.
slip Ceramic body containing 20 to 30 per cent water. Slip is clay in liquid form.
specific gravity Weight of a material compared to water.
sprig Method using a plaster mold to apply a relief design.
talc Ceramic material used in making wall tile. Talc is the softest rock in nature.
terrazzo Mixture of cement and marble chips to produce a colorful ceramic product.
thermal strengthening Heat-treatment process used to strengthen glass.
throwing Process of making ceramic objects on a wheel.
transparent See-through decorative or protective coating.
triaxial body Mixture of clay, flint, and feldspar in a number of proportions.
turning Process by which formed ware rotates past a cleaning tool to remove flaws.
underglazes Colored ceramic coating fired into ceramic bodies. Underglazes are porous and should be sealed by a glaze.
vitrifies Point at which ceramic ware hardens and becomes water resistant during firing.
ware Shaped or formed ceramic objects.
water content Percentage of water mixed with dry ceramic materials. Forming methods are classified by water content.
wedging Process used to make clay uniform, by removing air pockets.

Index

Abrading, 13, 14
Abrasives, 18, 107, 109
Aerospace missions, ceramics and, 111, 112
Aggregates, 98
Alkaline glaze, 73
Alumina, 72, 89, 98
Alumina clays, 9
Aluminosilicate glass, 90
Aluminosilicate refractories, 84
Antiquing, 75
Antistick ingredients, 43
Auger extruders, 39–42

Balances, 21
Ball clays, 8, 103
Bar mill, 13
Batch dryers, 50
Binder, 43
Bisque, 58, 60, 63
Blow-molding glass, 92, 95
Blunging, 22
Bone ash, 9
Bone china, 103, 104
Borax, 73
Borosilicate glass, 89–90
Bricks:
    common, 99–100
    face, 100
    fireclay, 84, 86, 87
    forming, 42
    silica, 84
Brittleness, 6
Building construction (*see* Ceramic products, in building construction)
Burn-off machine, 94, 95

Calcining, 98
Calcium carbonate, 73, 98
Calcium cement, 97
Carbide refractories, 85
Carbon refractories, 85
Castables, 85
Casting, 26–29
Cellular glass, 91–92
Cement, 18, 97–99
Ceramic bodies, 15–19
Ceramic industry, 3
Ceramic products, 97–113
    abrasives, 18, 107, 109
    aerospace missions and, 111, 112
    in building construction, 97–103
        bricks, 99–100
        cement, 18, 97–99
        concrete, 98–99
        plaster, 18, 99
        sanitary ware, 103
        tiles, 100–104
    cutting tools, 109
    electrical and electronic uses, 106–10
    in the home, 103–6, 113
    medical uses, 111
    metal, porcelain glazed, 111, 113
Ceramic raw materials (*see* Raw materials)
Ceramics:
    classification of, 3, 4
    defined, 1
    properties of, 3–8
Chemically strengthened glass, 88
China, 103, 104
China clay, 103
Chromite refractories, 85
Clay:
    classifications of, 8–9
    consistency of, 24, 26
    forming (*see* Forming)
    origin of, 2
    types of, 2
Cleaning greenware, 65–66
Clean-up tool, 65
Clear glazes, 74
Clinkers, 98
Coatings, 18, 111, 113
Coil forming, 31–32
Coke, 85
Colored glass, 90
Colored glazes, 74
Coloring ceramic bodies, 70–71
Color properties, 7
Comminution, 12–13
Common bricks, 99–100
Compressive strength, 4
Concrete, 98–99
Conduction dryers, 49, 50
Cone plaque, 60, 61
Consistency of clay, 24, 26
Continuous dryers, 50–51
Continuous kilns, 55–58
Convection dryers, 49, 50
Core of the earth, 1–2
Cornish stone, 9

118 INDEX

Corrosion, 82
Corundum, 84
Crackle glazes, 74
Cracks, 47–49, 83
Crushing raw materials, 12–13
Crust of the earth, 1–2
Crystal glazes, 74
Crystallized materials, 87
Cutting tools, 109

Decals, 78
Decorating (*see* Finishing)
Deflocculants, 23
Dense body, 53
Dental field, ceramics and, 111
Die, 41
Dipping, 76, 77
Dottling, 63
Downdraft kilns, 54, 55
Drainage tile, 100, 102
Drain casting, 27–29
Drawing glass, 94
Dry-clay bodies, 24, 25
Dry forming, 42–44
Drying, 45–51
    cracks and, 47–49
    stages in, 46–47
    types of dryers, 48–51
    warping and, 47

Earthenware, 103
Electrical and electronic use of ceramics, 106–10
Electrical properties, 6–7
Electric kilns, 58
Elevator kilns, 56, 57
Englobe coating, 70
Erosion, 2
Evaporation, 46
Extrusion, 39–42

Face bricks, 100
Feldspar, 9, 72
Ferrites, 107, 110
Fettling, 66
Fiber glass, 91, 105
Fiber optics, 107, 109
Filter presses, 24, 25
Finishing, 66–80
    decorations applied over glazes, 75–80
        lusters, 76
        metallics, 76
        methods of application, 76–80
        overglazes, 75–76
    glass, 94, 95
    methods applied to the body, 67–71
    nonfired stains, 75
    over-the-body methods, 71–75
Fireclay, 8–9, 100, 101

Fireclay refractories, 84, 86, 87
Firing: (*see also* Kilns)
    purpose of, 51–52
    types of, 61, 63
Fit, 73
Flaws, 65–66
Flint, 9–10, 16, 72 (*see also* Silica)
Flint clay, 8
Floor tiles, 102, 103
Flow properties, 7
Fluxes, 9, 16, 53, 72, 73, 75
Forming, 20–44
    casting, 26–29
    classification of methods, 20–21
    clay preparation, 21–26
        consistency, 24, 26
        dry bodies, 24, 25
        for machine molding, 33, 36, 37
        mixing, 22
        plastic bodies, 24, 25
        semidry bodies, 24, 25
        slip, 22–23
        weighing, 21–22
    dry, 42–44
    glass, 92–96
    hand-forming, 29–32
    hand and machine forming, 32–35
    machine molding, 36–42
        clay preparation, 33, 36, 37
        extrusion, 39–42
        jiggering, 36–38
        jolleying, 36–39
        rolling, 39
    refractories, 85
    semidry, 42–44
    stiff-plastic, 42
Frit glazes, 74
Froth flotation, 15, 16
Fusion forming, 85

Geology, 10
Glass, 16, 18, 87–96
    finishing, 94, 95
    forming, 92–96
    major types of, 88–90
    medical uses of, 111
    properties of, 87–88
    special, 90–93
Glass bodies, 18
Glass-ceramics, 92–93
Glazed structural tile, 100, 101
Glazes, 66
    firing, 60, 63, 73
    frit, 74
    ingredients of, 72
    metal, 111, 113
    surface appearance of, 74–75
    types of, 72–73
Glossy glazes, 74

Glost fire, 60, 63
Greenware, 60, 63
  cleaning, 65–66
  finishing (see Finishing)
Grinding raw materials, 13, 14
Grog, 75
Grout, 102
Guide cone, 60
Gum solution, 75
Gypsum, 10, 97–99

Hand-forming, 29–32
Hand and machine forming, 32–35
Hardness, 4–6
Heat-resistant glass, 105, 106
Hotel china, 103
Hot pressing, 43–44
Humidity, 47
Hydration, 98
Hydraulic mining, 12
Hydraulic presses, 42, 43
Hydrofluoric acid, 88

Igneous rock, 2
Impact strength, 5
Incising, 67
Inlaying, 67–68
Insulating firebrick, 84
Insulator, 6
Isostatic pressing, 43, 44

Jiggering, 36–38, 65
Jolleying, 36–39

Kaolin, 8, 72
Kick wheels, 33
Kilns, 51–63
  cement, 98
  construction of, 58–59
  early, 53–54
  firing cycle, 52–53
  heating, 58
  stacking, 63
  temperature measurement, 59–62
  types of, 54–58

Lava, 2
Lead-alkali glass, 89
Lead glazes, 73
Lead oxide, 72
Lead pencils, 105–6
Lehr, 96
Lime, 10, 97, 98
Limestone, 98
Liquid decorations, 76, 77
Lusters, 76

Machine molding, 36–42
  clay preparation, 33, 36, 37

  extrusion, 39–42
  jiggering, 36–38
  jolleying, 36–39
  rolling, 39
Magma, 2
Magnesia refractories, 84–85
Magnetic separation, 14–15
Mantle of the earth, 1–2
Matte glazes, 74
Maturing, 53
Mechanical properties, 4–6
Medical uses of ceramics, 111
Mesh size, 14, 15
Metal, porcelain coated, 111, 113
Metallic coatings, 76
Metamorphic rock, 2
Metric system, 21–22
Mining raw materials, 10–13
Mixing clay, 22
Moh's scale, 5
Mortar, 99
Mosaic tiles, 102
Muffle kilns, 54–55

Nepheline syenite, 9
96 per cent silica, 90
Nonfired stains, 75

Opaque underglazes, 71
Open-pit mines, 11, 12
Optical glass, 90
Optical properties, 88
Organic matter, removal of, 52–53
Overburden, 11
Overglazes, 75–76
Oxides, 72

Parison, 92, 95
Periodic kilns, 55–58
Phosphoric acid, 88
Photochromic glass, 90–91
Piercing, 67, 68
Pinch forming, 30–31
Piston-type extruders, 41, 42
Plaster, 18, 99
Plastic-clay bodies, 24, 25
Plastic clays, 9
Plastic forming:
  hand-forming, 29–32
  hand and machine forming, 32–35
  machine molding (see Machine molding)
Plastic forms, 85–87
Plasticity, 5–6
Pneumatic presses, 42, 43
Porcelain, 18, 53, 103, 111, 113
Porosity, 7, 66
Portland cement, 98
Potash, 73
Potters, 32–33, 46

INDEX

Potter's wheel, 32–35
Precast concrete forms, 99
Pressing glass, 93
Prospecting, 10–11
Protection, 66
Pug mills, 36, 37, 42
Pyrex, 105, 106
Pyrometers, 59–62
Pyrometric cone equivalent, 83
Pryometric cones, 59–62

Quartz, 9, 16, 72 (see also Silica)
Quartz glass, 90

Radiation dryers, 49–50
Raku glazing, 75
Raw materials:
   ceramic bodies, 15–19
   crushing and grinding, 12–14
   mining, 10–13
   origin and development of, 1–3
   refinement of, 14–16
   size classification, 14, 15
   types of, 8–10
Refractories, 18, 53, 72, 82–87
   available forms, 85–87
   classification of, 83–85
   defined, 82
   forming, 85
   properties of, 83
Reinforced concrete, 99
Relief decorations, 68–69
Residual clay, 2, 8
Roll crusher, 12, 13
Rolling, 39
Rolling glass, 94, 96
Rubber stamps, 77–78

Saggers, 55
Salt glazing, 74–75
Sanitary ware, 103
Scales, 21
Seam lines, 65–66
Sedimentary clay, 2, 8
Semidry-clay bodies, 24, 25
Semidry forming, 42–44
Semivitreous china, 103
Sgraffito, 70
Shaft mine, 11–13
Shrinkage, 7, 8, 45–46
Shuttle kilns, 56, 57
Silica, 9–10
   in carbides, 85
   in glass, 87–90
   in glazes, 72
   in Portland cement, 98
   refractories, 84
   in tiles, 100
Silk screening, 78, 94, 95

Slab forming, 31, 32
Slaking, 22
Slip, 7, 22–23, 47
   in casting, 26–29
   colored, 66
   removing excess, 65
Slip trailing, 70, 71
Slip welding, 31
Soda, 73
Soda ash, 23
Soda-lime glass, 88–89
Sodium silicate, 23
Sodium tannate, 23
Solid casting, 26–27
Spalling, 83, 85
Spare, 27
Specific gravity, 83
Splash mark, 66
Sponging, 66, 77
Spraying, 77
Sprig decorations, 70
Stains, nonfired, 75
Stiff-plastic forming, 42
Stoneware, 103
Stoneware clays, 9
Surface characteristics, 7, 42

Talc, 10
Tensile strength, 4, 5
Terrazzo, 99
Thermal conductivity, 6, 83
Thermal expansion, 83
Thermal shock resistance, 6
Thermal strengthening, 88
Thermometers, 59
Throwing, 32–35
Tiles, 100–105
Transfer paper, 78–80
Transparent underglazes, 71
Triaxial bodies, 16, 17
Tunnel dryers, 50–51
Tunnel kilns, 56–58
Turning, 66

Underglazes, 71–72
Underground mine, 11–13
Updraft kilns, 54

Vitreous chinaware, 103
Vitrifies, 51

Wall tiles, 102, 103, 105
Warping, 47
Waste chopper, 36, 37
Water content, defined, 20
Wedging, 29–30
Weighing ingredients, 21–22

Zirconium refractories, 85